ISLAND OF PARADISE

JONATHAN DOWNES

Crash Retrievals, Chupacabra, and Accelerated Evolution on the island of Puerto Rico

Typeset by Jonathan Downes,
Cover and Layout by M. L'orange for CFZ Communications
Using Microsoft Word 2000, Microsoft , Publisher 2000, Adobe Photoshop CS.

Photographs © 2008 CFZ except where noted

First published in Great Britain by CFZ Press

**CFZ Press
Myrtle Cottage
Woolsery
Bideford
North Devon
EX39 5QR**

© CFZ MMVIII

All rights reserved. Without limiting the rights under copyright reserved above, no part of this publication may be reproduced, stored in or introduced into a retrieval system, or transmitted, in any form of by any means (electronic, mechanical, photocopying, recording or otherwise), without the prior written permission of both the copyright owners and the publishers of this book.

ISBN: 978-1-905723-32-4

For Corinna

Acknowledgments

Nick Redfern for being my constant companion and partner in crime during the expedition; Richard Freeman for fulfilling the same role at home; Graham Inglis for following me round Central America, camera in hand back in 1998; Corinna (when I started this book, my girlfriend, now my wife) for typing the damn thing and providing cuddles and gin; Darren Naish for telling me that my theory wasn't nonsense; Spider McGraw for sitting on my knee and purring at the slightest opportunity (yes, folks he IS a cat); everyone at Cosgrove Meurer for their help and support; Norka, 'Cheese', Carola, and the crew; Judith Jaa`far for her UFOlogical input; my late father for his continued support; and above all my late mother for setting me on a life path when I always want to see just what is beyond that next turn in the road...

Foreword

So there I was in Woolfardisworthy, Devon – the village so nice they named it twice.

Known locally as the more tongue-friendly Woolsery, it's the sort of picture perfect setting favoured by Sunday night TV dramas, where murder is blended with cosiness. The kind of place where you expect the locals to be so inured to the discovery of corpses that they're casually stepping over them to reach their hanging baskets. But this was the real world, and my research had turned up no record of anything so exciting as a light happy-slapping – let alone a murder – ever happening in Woolsery.

It seemed to tick all the boxes of rural, pastoral cliché: twisting lanes walled by towering hedgerows, a Norman church with a well-stocked graveyard (natural causes, naturally), a heritage that could be traced back to the Domesday Book, and a young farmers society. The Woolsery year was punctuated by safe, murderless annual events, like the Woolsery Street Fair, the Woolfardisworthy Amateur Dramatics Society's Christmas panto, and The Weekend of Gardens, where locals invite one another to sniff their blooms without even the slightest threat of being garrotted in a greenhouse.

However, in recent years Woolsery's rustic idyll has become home to a group of people who – initially, at least – must have had John Constable choking on his own casket dust. And they were the reason I had driven down to Devon.

"You skinny bastard."

It was not a traditional greeting, and certainly not the one I'd been expecting, particularly given that it has been some decades since I was anything even approaching skinny (and a good few years since I was last a bastard), but everything is relative.

Island of Paradise

Jonathan Downes was one of the largest men I had ever met. Standing a good few inches above me – and I'm hardly a shrivelled Hobbit myself – Jonathan was almost as wide as he was tall. His face hidden behind an uncultivated beard of nicotine-hued wire wool, and framed by hair that apparently hadn't gotten a sniff of a brush in years – let alone been in scalping distance of a pair of scissors – he filled the room in every conceivable sense. The room in question was the characterful kitchen of the Centre for Fortean Zoology, the world's only full-time cryptozoological research organisation, and Jonathan Downes was a professional monster hunter.

He spoke in a voice that settled somewhere halfway between Noel Coward and Lesley Phillips, and I barely resisted the urge to add my own fruity "Ding dong" at the end of every sentence.

"Call me Jon, dear boy. All my friends do."

Quite how I ended up becoming friends with a group of professional monster hunters – let alone how I later embarked on a full-blown cryptozoological expedition to South America – is a story for another time. Suffice to say, monster hunts or not, as a screenwriter and retired video games journalist (occasionally writing under the ridiculous assumed name, 'Mr Biffo'), by trade I fall some way short of being able to call myself a cryptozoologist. What's more, I'm possibly more sceptical than most when it comes to matters of the unexplained. Certainly, if you're reading this book, it's likely that you're already playing Mulder to my Scully (though you can wipe that filthy thought out of your head right now, mister – you know full well what I mean).

Nevertheless, I have indeed become a friend of the Centre for Fortean Zoology. More than that – I'm a supporter. I have known Jon Downes for a year now, and in that time I have had my horizons broadened more than I could have imagined. I don't know what I'd expected when I accepted that invitation twelve-months ago, but I certainly wasn't expecting to have my world redefined.

We live in a society that uses the mundane to keep us all in check. Our imaginations have been neutered by telly, and fast food, and a media that ranks a knickerless heiress as more important than the collapse of our eco-system. Two point four kids. Nine to five. Sit down. Shut up. Do as you're told.

The CFZ dare to kick against all this. They dare to dig their heels in, and insist that there is another way. They state that the world has yet to give up its mysteries, and that there are answers, and questions, and wonders still waiting to be discovered.

It doesn't matter whether you believe in monsters, or mystery animals, or even aliens. As Jon Downes demonstrates in this book, the important thing is to open your mind to the *possibility*.

Ding dong indeed.

Paul Rose
May 2008

Part One
EXODUS

*"We sail tonight for Singapore,
we're all as mad as hatters here
I've fallen for a tawny Moor,
took off to the land of Nod
Drank with all the Chinamen,
walked the sewers of Paris
I danced along a coloured wind,
dangled from a rope of sand
You must say goodbye to me"*

Tom Waits/Kathleen Brennan `*Singapore*`

Prologue

It was early evening as the surprisingly tiny Boeing 757 flew steadily through the Caribbean dusk. Below us were the blue crystal waters, which in my childhood storybooks were populated by cannibal kings, fearsome pirates, and man-eating sharks. As I have never really grown up, my imagination still tended to view these waters in much the same manner as I had done in my pre-teens. However, as my Socio-political vistas broadened, and my favourite reading started to include Che Guevera [1], and to my eternal embarrassment, for a few months during my teens, Charles Berlitz [2], the heroes of the glorious workers and peasants counter-revolution, 'Papa' Doc Duvalier, and the Bermuda Triangle, joined Robinson Crusoe, Sir Francis Drake, Biggles, and the cast of *Treasure Island* in my mental seascape of the Caribbean oceans.

Today, however, as an overweight bearded zoologist, my daydreams went back only half a century to the days when another overweight bearded zoologist - my hero, Gerald Durrell - flew towards another South American airport, and described the delightfully *laissez faire* attitude of the locals aboard the plane. He wrote how this had been the only aeroplane on which he had travelled, on which he had had to share the cabin space with crates of chickens, and even a trussed-up pig, which were accompanying their owners on the journey.

I looked around me, and like my hero, took a hearty swig from a large tumbler of iced gin and tonic. In these less earthy days there were no chickens and no pig, but the rest of my co-passengers were remarka-

---

1. Ernesto 'Che' Guevara (1928-1967) was an Argentine born Marxist revolutionary who was one of the founding fathers of modern Cuba. He went on to fight in other Marxist revolutions across the world, and was killed in Bolivia. However he is best known, not for his 1960 manual on guerrilla warfare, or his book of his 1952 adventures, entitled *The Motorcycle Diaries* which has been described as *"Das Kapital* meets *Easy Rider"* but for the iconic image of him from a photograph by Alberto Korda which has been plastered over a bizarre range of merchandise for the last 40 years. The Marxist revolutionary has become an icon of cheapo cash in merchandising.
2. Charles Frambach Berlitz (1914-2003) was an internationally famous linguist, who moonlighted with books on fortean, and quasi fortean subjects such as *The Bermuda Triangle*,(1974) and *The Roswell Incident* (1980) as well as books on the search for Noah's Ark and the 'Philadelphia Experiment'. Much admired as a linguist (the People's Almanac once listed him as one of the fifteen most eminent linguists in the world), most of the more rational commentators on the world of fortean investigation consider his fortean writings to be a load of tosh!

bly familiar to me from Gerald's writings.

There was the big jolly black lady nearly as wide as she was tall, with her hair bound tightly in a brightly coloured bandana and surrounded by a gaggle of small children.

There was the old farmer returning to his homeland after a visit to relatives in New York.

There were the throngs of children and young people whose chatter was irresistibly reminiscent of the noises of the rainforest, which I would be hearing in only a few short days. The chirruping of cicadas, the raucous squeaking of the mountain tree frogs, and the delightful splishity splashity sound of the tiny mountain waterfalls.

But there were new additions to the human landscape before me. Additions, which I felt that perhaps my mentor would not have recognised, but which I think might have amused him. Sitting near the exit doors on the starboard side were three tall thin and grim looking 'Yardies' (well they looked like them anyway). Dark glasses, shoulder-length dreadlocks, and leather jackets, they glowered at the rest of the plane like predatory spiders perched in the corner of their web. A few seats away from them were four young Latino youths, whose t-shirts proclaimed that they were returning from a New York rappers' convention, joyously shouting rhyming couplets at each other, like tropical birds trading squawks in mid-air. Try as I might I could not make out the words - they were lost forever in the hubbub of humanity - but as I looked around me, and saw specimens from all walks of human life; the policeman handcuffed to his prisoner, the off-duty soldiers on furlough, and even the tall, painfully thin and very old man reading a German newspaper in the corner and looking for all the world like the popular image of a fugitive Nazi war criminal (when he was certainly just a respectable businessman on holiday), I realised that my mentor had been right; one can learn just as much about animal behaviour by watching an aircraft cabin full of people, as one can do in the most teeming jungle.

We landed at Luis Munoz Marin International Airport just as the last few glimmers of the tropical sun set below the horizon. Surprisingly, for an immigration lounge in an American controlled airport post-9/11, it was practically empty. On the mainland, everywhere you looked there would be policemen and military personnel, often brandishing machine guns, and always glowering at each and every one of the people who filed past them brandishing their passports as if each one was a potential terrorist. Here there was just a jolly fat black bloke in a uniform which reminded me irresistibly of that worn by the parkkeeper in one of the old *Beano* comic strips of the 1960s.

I proffered him my passport, and he just smiled. *"Shit man, you don't need to do that here"*, he grinned. I grinned back and clutching my hand luggage which contained my medication, cigarettes, a change of clothing, half a dozen cds, and a litre and a half of Jack Daniels, and what the bloke on the train the previous day had euphemistically described as a 'light picnic lunch' in a garishly coloured paper bag, I made my way to the baggage collection.

Suddenly it was as if I had been transported into the middle of a comedy funfair as designed by Hieronymus Bosch. It was chaos.

All the people that I had noticed on the aeroplane had converged upon the baggage collection console, and were alternately overjoyed to see their property again, or outraged when - like me - they were told that it had been diverted in transit to Anchorage, Alaska, or some likewise exotic location. The four-man rapping crew had no sooner collected their baggage when out came a beatbox, on came a complex pattern of resounding breakbeats, and they started to dance and sing. The enormous black lady with her hoard of children started to sway rythmically in time to the music, and before long the baggage console was deserted as everybody, including the Martin Bormann lookalike [1], and the three 'Yardies', decided to stop what they were doing, and watch the free cabaret.

A highly flustered looking individual wearing an American Airlines uniform started to take the flight details of everybody whose luggage had become mislaid, and when it came to my turn I gave my forwarding address as the *Windchimes Tavern* on Toft Street, San Juan, and oblivious of the fact that I was truly a stranger in a strange land - I had only one change of clothing, and nothing but a litre and a half of whisky, a British Rail pork pie, and $40 to sustain me, I made my way into the night.

The tropical night enveloped me like a blanket and it was like meeting an old and valued friend whom I had not seen for many years. It tasted like copper, and smelt like garlic, and within a few seconds the assorted hubbub of humanity which I had grown used to on the flight, and which I had likened to the sounds of the small creatures of the tropical night, were replaced with the real thing. I hailed the first available taxi and gave the driver my destination. He was a surprisingly fey young man, dressed in a yellow silk shirt, a straw hat of the type Carmen Miranda was wont to wear, and a pair of tiny Bermuda shorts. Whereas I have always prided myself on giving the appearance of being an English gentleman explorer, whilst on my travels I was more than a little taken aback when my new friend obviously misunderstood my motives for travelling to the fleshpots of San Juan. I reassured him that I didn't want to be shown a good time, that I had no interest in visiting "an absolutely darling little pub" where I was bound to meet lots of new "friends", and finally managed to persuade him to drive me to my destination. This he did with the minimum of good grace. I paid my fare, tipped him ten bucks that I could ill afford, and got out of the taxi. Here I felt even more of a stranger in a strange land. It was one of the more picturesque parts of San Juan which dated from the late colonial times, and the air was so humid and pungent that the tropical night caressed you like a fetid shroud. Slinging my holdall over one shoulder, I took a deep breath and strode towards the door marked 'Reception'.

It was locked.

I rang on the bell and waited. For several minutes absolutely nothing happened. Then, after what seemed like half an hour, but was probably only a few minutes, a nasal voice crackled through the small intercom speaker by the doorbell *"'ello"* it crooned. *"It ees not here!"* and the night fell silent again. Behind me I heard a strange noise. It sounded like something was strangling a cat about three inches away from my ear. Startled I looked around, but could see nothing but an eight foot tall specimen of the sort of shrub normally sold in UK garden centres at a fraction of the size for £20 a throw. I peered at it, and out of the corner of my eye, saw something moving away suddenly. I realised what it was; it was *el cocqui*; the tiny tree frog which is almost the unofficial guardian spirit of the island.

In the dusk on the other side of the street three feral-looking youths wearing totally unnecessary dark glasses, and festooned in heavy gold jewellery, eyed me up in a predatory manner. I began to feel wary, and pressed on the doorbell again. Once again the voice rang out *"'ello! It ees not here!"* and went silent, but when I pressed it for the third time, nothing happened at all, until a fat, jolly, little man with a curunculated face like a highly polished conker appeared from around the corner, grabbed me by the wrist and said *"Eet is not here, she is here!"* and pointed to another door just around the corner. I then realised that what I thought were the outside walls of the hotel, were actually the walls to a large - and ever so slightly fortified - courtyard in which the hotel complex nestled. I asked my new friend where I had to go to sign in. To my surprise he knew my name already, and was also aware that my luggage had gone missing. He smiled at me affectionately and pointed me to a twisty little path that led by a swim-

---

[1]. Or to be more honest, the 'version' of Martin Bormann who appeared in the Sex Pistols movie *The Great Rock & Roll Swindle*. Only thinner (and more psychopathic)

ming pool towards a small archway. *"Go through there, Senor Jon"*, he said. By this time, completely confused, I did as I was told, and found myself in a small, but well-appointed courtyard bar. To my surprise, in the corner, sitting there as if they had been there for hours, were the three 'Yardies' from the airport, and the elderly Teutonic looking man whom I had likened to Martin Bormann. How they had got to the hotel and settled in whilst I was still trying to make sense of the concierge I had no idea. I smiled at them, and they waved at me with delighted grins on their faces as if we had known each other for years.

It was a strange looking bar, although in some ways it looked like the archetypal tourist trap, with plastic fruit, coconut palms and salsa music, it also had a weirdly surreal atmosphere about it, as if it were located in some sort of limbo where the normal laws of the space-time continuum didn't really hold sway. The clientele were typically atypical. Apart from my acquaintances from the flight (and you must remember that I hadn't actually spoken to any of them) there were only two; one was a man wearing a Hawaiian shirt and Bermuda shorts, with distinguished collar length silver flowing locks, and the other was a sinister looking character dressed head to foot in black, with a bandana emblazoned with skulls and crossbones wrapped around his shaven head. He looked the sort of person that I had spent my life trying to avoid, but I went up to him and tapped him on the shoulder.

"Mr. Redfern I presume", I said

"Well, I'll go to the foot of our stairs, it's our Johnny loike", the skeletal figure said in a broad Brummie accent. *"Bostin! I didn't think you were arriving until tomorrow, have a drink"*....

The Margarita Diary,

Prologue Comments:

When Jon asked me if I would summarise my own thoughts and recollections at the end of those chapters of *Island of Paradise* in which I played a role, I thought: well, why the hell not? After all, it's not every day that a player in someone else's book gets the opportunity to insert their own thoughts, observations, memories and recollections into the tale.

And so, here you have my first entry in *The Margarita Diary* (yes, I know: an utterly shameless play on the mighty Hunter S. Thompson's fabulous *The Rum Diary* that told of his own journalistic adventures on Puerto Rico back in the 1950s): a brief and concise commentary on that week in 2004 when Jon and I roamed the wilds of Puerto Rico in search of diabolical entities from the outer edge.

Reading Jon's prologue to *Island of Paradise* instantly brought back fun and fond memories of my own flight to Puerto Rico, where Jon and I met up to do battle with (well, okay, to search for) the vampire-like chupacabras. This was the first time that I had travelled to the island to seek out the mystery beast; and for me the build-up was one filled with both excitement and anticipation.

It had been far too long since Jon and I had roamed the woods, fields and villages of an exotic location in search of all-things monstrous and macabre; and, therefore, it was high time to rectify the situation.

I have to confess, however, that I was not at all prepared for the absolute wall-of-heat that hit me as I exited the doors of San Juan's airport and made my way to my driver, who was waiting to take me to whatever it was that fate had in store.

As dear Jonny gleefully likes to point out, I am a pale and pasty Brit of scrawny build and invariably black-garb; and, therefore, one who by definition does not fare at all well in climates pushing on 100 degrees or more. And, indeed, that much became abundantly evident as I fought my way through the crowded exit doors of the airport, and as the pummelling heat hit me squarely in the face like a rock.

'Bloody hell,' I thought, 'we don't have temperatures like this back in Birmingham,' - in reference to my home city back in the cooler climes of jolly old England, where I resided until midway through 2001.

Nevertheless, I was a man on a mission, and I was not about to let a bit of sun get the better of me. There were monsters to be found, old friendships to be rekindled, new ones to be made, and adventures to be had. And it all began over margaritas and dinner at the *Windchimes* hotel. The game was afoot, as a certain fictional seeker of all-things mysterious was so fond of saying... **NR**

Chapter One
The Island of Paradise

Puerto Rico was "discovered" in November 1493 by Christopher Columbus who was on his second voyage of discovery of the New World. When I was a little boy in Hong Kong, one of my favourite books was a *Ladybird* book which told an exciting story how the gallant little three-vessel fleet, financed by the bountiful and generous Queen Isabella of Spain, had taken the intrepid Italian explorer across the Atlantic Ocean in search of a new route to India. In easy bite-size chunks of text and attractive water-colour paintings it told Columbus' story and the climax with a picture of the brave explorer (looking a bit like Johnny Depp) being greeted by three beautiful subservient, semi-naked natives, giving him gifts of jewellery and looking ever-so pleased at their good fortune at having been "discovered" by the forces of imperial Europe.

What it didn't tell was the real story; a disgusting parade of violence, exploitation, murder, torture and inhumanity which resulted - only a few centuries later - in the original inhabitants of the region, the Taino Indians, becoming functionally extinct. From a population of more than 30,000 before the arrival of the Europeans, missionary reports of 1515 recorded only 4,000.

Michele de Cuneo, a Ligurian nobleman on Columbus' second voyage, wrote in 1495:

"While I was in the boat I captured a very beautiful Carib woman the said Lord Admiral gave to me, and with whom, having taken her into my cabin, she being naked according to their custom, I conceived desire to take pleasure. I wanted to put my desire into execution but she did not want it and treated me with her fingernails in such a manner that I wished I had never begun. But seeing that (to tell you the end of it all), I took a rope and thrashed her well, for which she raised such unheard of screams that you would not have believed your ears. Finally we came to an agreement in such manner that I can tell you that she seemed to have been brought up in a school of harlots."

It was a symbolic beginning to centuries during which the Island of Paradise and her peoples were to be raped by foreign oppressors. And as is so often the way, even now, the foreign oppressors seem to believe that their rape is no such thing, but a boon being granted by a beneficent, and horridly avuncular master-race to a sadly underprivileged cousin.

Columbus first established his base of operations on the island of Hispaniola (nowadays divided into Haiti and the Dominican Republic) but soon colonised the other islands in the region.

With the colonisation process, the first *encomienda* settlements on the island were established, in which lands and their indigenous peoples were allotted to the conquistadores. Under the *encomienda* system, conquistadors were able to use the native people as labourers and in return provide them with military protection. However, many conquistadors abused their power, and enslaved the native people. The Spanish Crown, under pressure from the Church, instituted the *repartimiento* system, under which the conquistadors were to pay the Indians for their labour, and to teach them Christianity. The system didn't last, and the Tainos were forced into slavery.

In 1508, Juan Ponce de León - a conquistador who had sailed with Columbus on his second voyage - founded the first settlement in Puerto Rico, Caparra (later relocated to San Juan). He was greeted with open arms by the Taino Cacique, Agüeybaná, and quickly gained control of the island. As a result, Ponce de León was named Governor of Puerto Rico in 1509.

Agüeybaná, whose name means "The Great Sun", was the principal "Cacique" (Chief) of the Tainos and the most powerful governing Taino in "Boriken" (Puerto Rico) when the Spaniards first arrived. He lived with his tribe in "Guaynia" (Guayanilla) located by the river of the same name, on the southern part of the island. All the other Caciques were subject to - and had to obey - Agüeybaná, even though they governed their own tribes.

Agüeybaná, believing that the Spaniards were Gods, received the Spanish conquistador Juan Ponce de León with open arms upon his arrival in 1508. According to an old Taíno tradition, Agüeybaná became "guaitiao" (friends) with Ponce de León, and their wives exchanged names. Ponce de León baptised the Cacique's mother into the Christian religion, and renamed her Inés. The hospitality and friendly treatment that the Spaniards received from Agüeybaná made it easy for them to conquer the island.

Ponce de León is an enigmatic, but oddly fascinating historical character. He was an explorer and mystic, administrator and adventurer, and he was a horribly cruel, though brave, man. The European colonial empires sometimes produced such men, who - on the one side - were capable of bestial acts of cruelty and viciousness, but who on the other hand appeared to have a very real empathy with the peoples that they so brutally subjugated. But Ponce de León - like so many other Colonial administrators - had an obsession. It was an obsession that would over-ride all vestiges of taste and decency in his nature, and it was an obsession that would eventually destroy him.

For centuries travellers had returned to Europe with stories of mysterious 'fountains of youth'; life-giving waters with apparently miraculous powers of healing and rejuvenation. The history of such stories is particularly interesting.

A 12th Century book of travels by a writer claiming to be Sir John Mandeville, (or Jehan de Mandeville) mentions a fountain of youth. The book was published in Anglo-Norman French between 1357 and 1371. By aid of translations into many other languages it acquired extraordinary popularity. Despite the extremely unreliable and often fantastical nature of the travels it describes, it was used as a work of reference by explorers as recently as Ponce de León's mentor, Christopher Columbus.

In it, Mandeville claims to have visited a city called Polombe, which many contemporary experts believe was situated roughly where the present-day city of Kollam, a city and a municipal corporation in Kollam

JUAN PONCE DE LEON.

Agueybana greeting Ponce de León upon his arrival in Puerto Rico
BELOW: The Kerait ruler Wang Khan depicted as Prester John in *Le livre des Marseilles* (15th Century)

district in the Indian state of Kerala. It is an old sea port town. Kollam had a sustained commercial reputation from the days of the Phoenicians and the Romans. Pliny (23-79 AD) mentions about Greek ships anchored at Musiris and Nelkanda. Musiris is identified with Kodungallur (then ruled by the Chera kingdom) and Nelkanda with Quilon or Kollam (then under the Pandyan rule). Spices, pearls, diamonds and silk were exported to Rome from these two ports on the South Western coast of India. Of these, pearls and diamonds came from Ceylon and the South eastern coast of India, then known as the Pandyan kingdom.

Whilst there Mandeville claims:

"at the foot of that mount is a fair well and a great, that hath odour and savour of all spices. And at every hour of the day he changeth his odour and his savour diversely. And whoso drinketh three times fasting of that water of that well he is whole of all manner sickness that he hath. And they that dwell there and drink often of that well they never have sickness; and they seem always young. I have drunken thereof three or four sithes, and yet, methinketh, I fare the better. Some men clepe it the well of youth. For they that often drink thereof seem always young-like, and live without sickness. And men say, that that well cometh out of Paradise, and therefore it is so virtuous."

Unfortunately, most of the available evidence suggests that large parts of the book are mere invention. There is no known contemporary corroboration of the existence of Jehan de Mandeville for example. Some French manuscripts, not contemporary, give a Latin letter of presentation from him to Edward III, but it is so vague that it might have been penned by any writer on any subject. It is in fact beyond reasonable doubt that the travels were in large part compiled by a Liège physician, known as Johains a le Barbe or Jehan a la Barbe, otherwise Jehan de Bourgogne. However, some of the source material upon which he drew probably *was* genuine, and the stories of the fountain of youth were very prevalent in traveller's tales for several hundred years. [1]

In about 1165, copies of a manuscript entitled the *Letter of Prester John* started spreading throughout Europe. It was supposedly written to the Byzantine Emperor Manuel I Comnenus (1143 – 1180) by Prester John, descendant of one of the Three Magi and King of India.

Prester John was allegedly the monarch of a lost kingdom of Christians somewhere in the vastness of Central Asia. The Kingdom of Prester John was associated by many scholars with St. Thomas the Apostle - best known as 'Doubting Thomas'; the one of Christ's disciples who only believed that Our Lord had risen from the dead, after he physically examined the wounds on His hands and feet. After Pentecost, Thomas travelled to the east, and spread the Gospel to the Parthian Kingdom - an Iranian civilization situated in the northeast of modern Iran, but at its height covering all of Iran proper, as well as regions of the modern countries of Armenia, Iraq, Georgia, eastern Turkey, eastern Syria, Turkmenistan, Afghanistan, Tajikistan, Pakistan, Kuwait, the Persian Gulf coast of Saudi Arabia, Bahrain, Qatar, and the UAE.

He also, allegedly, travelled even further east. Evidence to support this comes in a hymn by an obscure Saint called Ephrem the Syrian (died 373) who includes the following lines, in which the Devil cries:

"...Into what land shall I fly from the just?
I stirred up Death the Apostles to slay,

---

1. The biographical information (if biographical is the right term to use for people who never existed) for Prester John, and the elusive Sir John M comes from various sources including Wikipedia, Sabine Baring-Gould's *Myths of the Middle Ages*, (1876) and an invaluable tome called *Animal Fakes and Frauds* (1978) both of which I heartily recommend to anyone. However, these characters are fairly tangental to the main thrust of the story so a full bibliography would be superfluous.

> that by their death I might escape their blows.
> But harder still am I now stricken:
> the Apostle I slew in India has overtaken me in Edessa;
> here and there he is all himself.
> There went I, and there was he:
> here and there to my grief I find him".

This hymn was published by Fr. Medlycott, the founder of St. Thomas College, Thrissur, the Cultural Capital of Kerala State.

Wikipedia, the free encyclopaedia, says:

"It is believed that St. Thomas had come to Kerala, India to spread Christianity. Even today people flock to the Church at Malayatoor. He further moved towards north by coast and reached a small village called Palayur, near Guruvayoor which was a priestly class settlement at that time. Here he conversed with priestly class community of Aryan worship system. Convinced by the Divine power possessed by this foreign monk of new faith, four prominent rich and priestly class families accepted the Christian faith and were baptised by St.Thomas himself. The four prominent high class priestly Hindu families who accepted the new faith were Kali, Kalikavu(Kaliyankal), Pakalomattom and Sankarapuri."

Against this background of unconscious quasi-collaboration, it is no wonder that so many people across Europe believed implicitly in the literal truth of the Kingdom of Prester John. [1] The ubiquitous 'letter' included the following passage:

"At the foot of Mount Olympus bubbles up a spring which changes its flavour hour by hour, night and day, and the spring is scarcely three days' journey from Paradise, out of which Adam was driven. If anyone has tasted thrice of the fountain, from that day he will feel no fatigue, but will, as long as he lives, be as a man of thirty years. Here are found the small stones called Nudiosi, which, if borne about the body, prevent the sight from waxing feeble and restore it where it is lost. The more the stone is looked at, the keener becomes the sight."

The 'fountain of youth' also crops up in *The Alexander Romance* - several collections of legends concerning the mythical exploits of Alexander the Great. This was another bestseller in 12th Century Europe, although the earliest versions date to 3rd Century Greece, even *then*, some 600 years after Alexander's death at the age of 33.

Having been given spurious endorsement by this motley collection of somewhat dubious sources, the hunt for the 'fountain of youth' became one of the most important priorities for the Catholic Church, and therefore for successive generations of European explorers.

At the time that the Spanish first arrived in the Caribbean, the location of the fountain was generally believed to be somewhere in eastern Asia, and they all told of a famous spring - the waters of which had the marvellous virtue of restoring to youth and vigour those who drank them. The Spaniards heard from the Indians of Puerto Rico tales of magickal life-giving waters that reminded them of this *Fons Juventutis*,

---

1. There is an irresistible parallel between the ubiquitous spread of such 'knowledge' in the middle ages, and the way that the twin Gospels of - for example - Che Guevara and Charles Berlitz whom we discussed in the footnote of page 13 were spread amongst those of my generation in the late 1970s. The truth is, of course, that although Nelson Mandela referred to Guevara as *"an inspiration for every human who loves freedom"*, Johann Hari wrote that *"he was an actual person who supported an actual system of tyranny"*. Guevara was a brutal killer, and Berlitz, for much of the time at least, wrote pernicious nonsense, but they both became icons to a generation. One of the reasons that I quote from *Wikipedia* in this book to such an extent is that I am fascinated by the concept of a 'consensus reality'; a universe where 'truth' is decided by committee. In many ways, the decidedly dodgy concept of 'truth', and the even dodgier one of *"..but everybody knows that".*. is the biggest central theme of this book

and in most of the history books written about him, it is acknowledged that for much of the rest of his life Ponce de León searched in vain for them. Indeed, the very name of Juan Ponce de León has become synonymous in popular culture with the search for the fountain of youth. But according to local legend, the story is more complex than this.

In my writings I try to provide proper references for my revelations, especially when they are - as I believe this one is - of some little importance. However, I was told this story in the bar of our hotel by an elderly schoolteacher whom I know only as 'Geraldo', and so I cannot provide any corroboration at all, and I am painfully aware that *'pers.com from some bloke I met in a bar'* does not really hold much water as an academic reference. [1]

However, according to Geraldo - as we have seen - during his early times on the island Ponce de León was initally greeted as a friend by the locals, and indeed had made quite close friendships, most notably with the island's tribal leaders. Ponce de León was somewhat of a mystic, and was interested in the stories that he was told about the mysterious life-giving qualities of some of the springs on the mountain now known as El Yunque. These waters were sacred, indeed - as I was to discover - they were revered as divine waters, and Ponce de León is said to have considered himself to be very privileged to have been told about them.

But, the honeymoon period didn't last very long.

By 1511, a year after Agüeybaná's death, the Tainos began to revolt against the Spanish settlers. It is believed that the chief ordered his warriors to drown Diego Salcedo, a Spanish bishop, to determine whether or not the Spaniards were immortal, as they believed that Spanish colonisers had divine powers. After they drowned Diego, they kept watch over his body for several days until they were sure that he was dead. The Tainos, after learning that their captors were not immortal, revolted against Spanish rule. The revolt was easily crushed by Juan Ponce de León, who ordered 6,000 shot; and the survivors fled to the mountains, or left the island.

According to Geraldo, Ponce de León's erstwhile friends amongst the local shamanic community, told him that as a result of this massacre, he was cursed, and would surely die if he were to continue his hunt for the *Fons Juventutis* in the Puerto Rican mountains.

So Ponce de León was forced to take his search elsewhere.

The Tainos had told the Spanish of a large, rich island to the north named Bimini, and Ponce de León was searching for gold, slaves and lands to claim and govern for Spain, all of which he hoped to find at Bimini and other islands. Ponce de León equipped three ships at his own expense, and set out on his voyage of discovery and conquest in 1513. On March 27, 1513, he sighted an island, but sailed on without landing. On April 2 he landed on the east coast of the newly "discovered" land at a point which is disputed, but was somewhere on the northeast coast of the present State of Florida. Ponce de León claimed "La Florida" for Spain. He named the land *La Florida*, meaning *flowery*, either because of the vegetation in bloom he saw there, or because he landed there during *Pascua Florida*, Spanish for *Flowery Passover*,

---

1. However, much of it has been borne out by what is in the public domain about Ponce de León in the public domain. For those of you interested in finding out more, I refer you to the following titles:

Rivers of Gold: The rise of the Spanish Empire by Hugh Thomas (Wedenfield and Nicholson, London, 2003)
Juan Ponce de León: The exploration of Florida and the search for the Fountain of Youth by Robert Greenbeger (Rosen Publishing, 2003)

meaning the Easter season. *Pascua Florida Day*, April 2, is a legal holiday in Florida. [1]

Ponce de León then sailed south along the Florida coast, charting the rivers he found, passed around the Florida Keys, and up the west coast of Florida to Cape Romano. He sailed back south to Havana, and then up to Florida again, stopping at the Bay of *Chequesta* (Biscayne Bay) before returning to Puerto Rico. But, ironically, Ponce de León may not have been the first European to reach Florida. He encountered at least one Indian in Florida in 1513 who could already speak Spanish.

---

1. This is where the story really begins to get peculiar. Over the years I have given up becoming surprised at seemingly ridiculous coincidences. Indeed my old friend and mentor Tony Shiels once said to me that there is no such thing as a coincidence, and I think he's probably right. British explorer Colonel John Blashford Snell and I have known each other for nearly 20 years. In 2001 he generously did us the honour of becoming Life President of the CFZ, and every Christmas or thereabouts he telephones for a chat. 2008 was only a few days old when he called me, and asked what projects I was working on. As I was putting the final touches to the last couple of chapters of this book I told him about it and he burst out laughing. *"You know that I went searching for Ponce de Leon's fountain of youth, don't you?"* he asked.

The following passage is from his 1983 book *Mysteries* (Bodley Head), and is used with his permission:

Nevertheless, there was still a mystery to investigate. A spring of apparently fresh water, and located in the mangrove near Bone Fish Hole, North Bimini, was, according to popular legend, a possible site of the Fountain of Youth which the Spanish explorer Juan Ponce de Leon, may have been seeking when he discovered Florida in 1513. It was believed that to bathe in the spring would ensure perpetual youth, and there was a rumour that the waters contained certain chemicals. I wondered if the might be a link between this legend and the story of Atlantis. So it was, on a relatively cool but thundery morning, that we piled equipment into Les Savage's Zodiac, and set off across the shallow lagoon that is the centre of Bimini. As we edged over the bars, our propeller grinding through the sand, large black shadows glided away - the place was simply crawling with sharks.

Although the site was fairly well known and I'd seen it from the air, it took the ascent of a mangrove tree by Clark Shimeall finally to locate the spot. To reach the spring we had to push the boat into a narrow channel which ran for 50 meters through the mangrove swamp. The thunder rumbled and rain began to fall in heavy spots as we approached the 'Fountain of Youth'. Eager to be the first there, I pushed ahead, wading in the shallow water and holding aloft my cine camera and the sterile specimen bottles we should need to sample the magical water. Suddenly, the channel opened out to reveal the spring - in fact, a deep pool, but before I could utter a word my feet sank deep into quicksand and I plunged, cameras and all, up to my neck in the water.'oh well! Let's hope it is the fountain`, said my colleague, and pulling on our face masks, we plunged into the spring. The well -if that is what it was - measured around 10 feet deep, and was some 38 feet in diameter. Underwater it was as clear as crystal. Dark snake like mangrove roots protruded from the sides, but the water was quite fresh to drink start the sort of several middle-aged men diving around in a supposed Fountain of Youth seemed faintly ridiculous.

It is rumoured that the water of this well contained lithium, so out came the specimen bottles and samples were taken from various depths. Lithium is used for treating various psychiatric disorders and maybe something of a tranquilliser, so perhaps this was the reason for the legend stopped however, although we felt tired the visit and on return [...] I slept for almost two hours, this may have been due to sheer exhaustion.

Back in Britain, the Canon Company once again cleaned up my camera. "'Fountain of Youth` seemed to have done it good!" They commented, and water samples were tested in Miami and Berlin, but, alas, there was no more lithium that you would expect in any water, nor indeed any unusual constituents.

The weird thing wasn't that the Colonel and I had both been looking for the Fountain of Youth, but that we had both been drawn towards it whilst searching for something completely different. As I was to find out, the search for the truth about the Fountain of Youth was to be a very strange one...

The death by forcible drowning of Bishop Diego Salcedo is a remarkably popular subject for sculptures even today

The last resting place of a cruel, but fascinating, man

Island of Paradise

In 1514 Ponce de León returned to Spain and received commissions to conquer the Caribs of Guadalupe, and to colonize the *Island of Florida*. His expedition to Guadalupe in 1515 was not successful, and he returned to Puerto Rico where he stayed until 1521.

In 1521 Ponce de León organised a colonising expedition on two ships. It consisted of some 200 men, including priests, farmers and artisans, 50 horses and other domestic animals, and farming implements. The expedition landed on the southwest coast of Florida, somewhere in the vicinity of the Caloosahatchee River or Charlotte Harbor. The colonists were soon attacked by Calusas, and Ponce de León was injured by a poisoned arrow. After this attack, he and colonists sailed to Havana, Cuba, where he died. His tomb is in the cathedral in Old San Juan.

The hunt for the truth behind the legends of Ponce de León is complicated by the fact that there were actually several different men with the same, or a similar, name. These included:

- **Juan Ponce de León II** (1570 – 1600), the first Puerto Rican to assume (temporarily) the governorship of Puerto Rico. He was the grandson of 'our' Ponce de León
- **Luis Ponce de León** (1461 – 1526), Spanish governor of New Spain
- **Luis Ponce de León** (1527 – 1591), Spanish poet, Catholic mystic, and victim of the Inquisition
- **Diego de Vargas Zapata y Luján Ponce de León y Contreras** (1643 – 1704) Spanish Governor of the New Spain territory of Santa Fe de Nuevo México, today the U.S. states of New Mexico and Arizona

I have found a number of instances where the lives and exploits of these men have become confused. The poet, especially, was a great mystic and spiritual writer, and sadly for us 21st Century seekers after truth, a number of the sources on which I have drawn whilst preparing this book have wrongly attributed his writings to the conqueror of Puerto Rico.

Puerto Rico became the hub of the Spanish Empire in the New World and once its original dusky inhabitants had been decimated and finally wiped out by slavery, war, influenza and syphilis it became one of first ports of call for the burgeoning slave trade in which luckless inhabitants of West Africa were captured, shipped to the New World and sold as cheap labour. The British Empire abolished slavery in 1833, and much of Britain's colonial expansion over the rest of the 19th Century was as a result of widespread British revulsion with the trade, and British attempts to finally wipe it out. The United States finally abolished slavery after the long and bloody Civil War, but the Spanish Empire continued trafficking in human flesh for many years. On March 22, 1873, the Spanish National Assembly - with very bad grace - finally abolished slavery in Puerto Rico. The owners were compensated with 35 million pesetas per slave, and slaves were required to continue working for three more years.

By 1825 the mighty Spanish empire was in disarray, and many former colonies had been given autonomy or even full independence. In the Caribbean, the only two Spanish possessions left were Cuba and Puerto Rico. In the 30 years following the American Civil War, much of the United States Foreign Policy had been to consolidate their national borders and to establish offshore naval bases. In the aftermath of the war with Mexico, which had created national heroes of the not very heroic - Davey Crockett - and the morally dubious - Jim Bowie - the United States became very Anglocentric; an irony considering that it had only been a century since the War of Independence. There was a great deal of national resentment towards those Americans of Hispanic origins, and moves were set in place to colonise the states which had been taken from the Mexican Empire with Americans of English descent. Ironically it could well be argued that in doing this, the Americans were mirroring the actions of successive British governments in Ireland that they were to condemn so vociferously over the next century.

By 1867 Puerto Rico had reached a population of 656,328; its population recorded as 346,437 whites and 309,891 "of colour", which included Blacks, *Mulattos* (people of mixed European and African origin) and *Mestizos* (people of mixed European and Tanio heritage) and the majority of islanders were living in extreme poverty. The agricultural industry, the main source of income, was hampered by lack of roads, rudimentary tools and equipment, and natural disasters, such as hurricanes and periods of drought. The intellectual minority remained relatively active within the limitations imposed by local Spanish authorities, and illiteracy was 83.7%. [1]

In the last quarter of the 19th Century there was a strong movement for independence in Cuba and, not for the last time, America backed the Cuban rebels. When in 1898 an American battleship was sunk in Havana Harbour the stage was set for the Spanish American War which, once and for all, secured American naval bases at Guantanamo Bay in Cuba and on Puerto Rico.

Puerto Rico, together with the Virgin Islands, Guam, the Philippines, and a few other bits and bobs of island real estate moved to being under American rule. They have been there ever since.

Puerto Rico is a strange place. Although it has been an American protectorate for over a hundred years, even now it has far more links with Spain than it has with the Land of the Free. I read up quite widely on its history and economics when I was writing my first book on the island, *Only Fools and Goatsuckers* (2000). However, as I discovered during my second trip to the island in 2004, for a number of reasons I had gained a largely unrealistic view of the island's infrastructure. This was basically because 1998, when I first visited was - as we shall see - the height of chupacabra fever in Latin America. Ours was only one of at least a dozen foreign film crews who had come along to the island, behaved badly, and disappeared again, although, as my first book showed, the resulting film *The Fearless Vampire Hunters* (UK Channel 4 May 1998) was quite sympathetic to the islanders and - to my chagrin - spent most of its time trying to make fun of Graham Inglis and myself, we had not engaged the sympathy of the islanders, who were minded to see us as *locos gringos* who were more than likely planning, like Muriel the Actor, to take the almighty piss [2], rather than present a balanced and scientific viewpoint of the phenomenon we had come to film. This time everything was different. We had an almost entirely local crew, and Nick Redfern and I endeared ourselves to them within hours by mucking in, trying to speak the language, joining in their jokes, and buying our round when it was our turn. We soon became accepted as scientists with a *genuine* interest in what had been happening on the island, and so, almost from the beginning, I found that many of my preconceptions made on my previous visit had been completely wrong.

Socially at any rate, Puerto Rico is not a province of the United States. If anything, spiritually at least, it is still a province of Spain, and it was interesting to find out that over a century after the Spanish Empire had been ignominiously routed, the more well to do inhabitants of the island still send their children to spend a few terms at a Spanish school wherever possible. It is interesting to compare the utter Hispanicisation of the island and its culture over a century after Spanish rule ended with what little can be seen of British Imperial rule only a few decades after we left our possessions in India and Africa.

The difference was, of course, that Puerto Rico, Cuba and to a certain extent Mexico, were truly colonies in the true sense of the word. Most of the British Empire wasn't. In my earlier book I drew parallels between Puerto Rico and the Crown Colony of Hong Kong where I had spent my childhood. It would perhaps be more apposite to compare Puerto Rico with Australia, New Zealand or Canada; places where

---

1. These facts and figures come from a number of sources, including Nancy Morris (1995), *Puerto Rico: Culture, Politics, and Identity*, Praeger/Greenwood, ISBN 0275952282, and of course Wikipedia which pinched most of its entry from the aforementioned Ms Morris.
2. *The Human Menagerie* by *Cockney Rebel* (EMI, 1973) for those who didn't get the allusion

Island of Paradise

the vast majority of the population is descended from the colonial settlers, rather than the indigenous peoples. However, as recent events on the political stage have shown, there is not much love lost between many Australians and New Zealanders and their country of ancestry, so I would love to know why the present day inhabitants of Puerto Rico feel such a strong bond with the European country of their Imperial past.

From a fortean point of view, Puerto Rico is particularly interesting [1]. It has widely been reported that during the second journey to the New World, Columbus had at least two UFO sightings. He is reported to have seen a fiery object speed through the sky and drop into the ocean near one of the islands in the Lesser Antilles, but it is also reported that he was loathe to tell his companions of this sight, fearing that they would take it as a bad omen. He is also reported to have seen strange ivory coloured lights in the sky above the islands of what we now know to be Bermuda. Christopher Columbus and Pedro Gutierrez - on the deck of the *Santa Maria*, - observed, *"a light glimmering at a great distance."* It vanished and reappeared several times during the night, moving up and down, *"in sudden and passing gleams."* It was sighted 4 hours before land was sighted, and taken by Columbus as a sign *"they would soon come to land."*

The UFO sightings have continued apace during the intervening five centuries.

There have even been accounts of UFO crashes, such as this one from May 1997, which was posted on Jeff Rense's website:

In Lajas [SW of the island], one of the places with most sightings in Puerto Rico," at around 9:00 PM, a big and luminous object was seen falling by many people followed by a loud noise. People from the town ran to the mountain to see what happened and witnessed an area of about one kilometer burned, big and small animals dead and whatever remained of the grass unburned crushed down as if a big roll crushed everything. There were a very foul smell [sic]like sulfur. The stench was so intense many had to return as they could not stand it.

The men saw a big and luminous green light and tried to reach it but it was on a very deep valley. It was around 11:00 PM and it was very dark when they decided to go back to town to seek for help, ropes, lanterns. When they came back the police stopped them.

During late night and early morning, the police was running around all over the place. First they said that there was a motorcycle, HA! HA! As if motorcycles fly, then they changed the story to a meteorite. When asked what the FBI had to do there with a meteorite and why the local police was kicked out from the area and why the presence of helicopters and the National Guard, they had no answer.

Hours before this, residents reported UFOs in the area and called the police. Some reports came from other parts of the Island. A woman in Ponce videotaped the bright light flying over a hill.

Everything has been covered up. There is no evidence. Nothing happened. Now only stands the memory of the fall of star and the grass burned and crushed. They are very efficient my friend. A unit specialized with experience.

Ponce, by the way, is a town in the south of the island, named after Ponce de León.

---

1. I refer the interested reader to two magnificent books by Scott Corrales: *Flashpoint:- High Strangeness in Puerto Rico* (Amarna Ltd , 1998) and *Chupacabras and Other Mysteries* (BookWorld Press,U.S., 1997). Both of these books are peerless, and tell far more about the fortean history of this remarkable island than I could ever do.

Puerto Rico is also one of the corners of the so-called Bermuda Triangle (also known as Devil's Triangle) - an area in the Atlantic Ocean which lies between the islands of Puerto Rico and Bermuda and the coast off of Miami, Florida in the United States. Rather than describing a passageway between wonderful vacation spots, the Bermuda Triangle is known as the place where ships at sea mysteriously disappear. The 'triangle' has also produced its share of UFO reports, and also some strange accounts of fast moving undersea objects. In 1963 over the Puerto Rico Trench, a Naval task force of helicopters and aircraft from the carrier Wasp and several smaller surface ships tracked and hunted an undersea object that had been detected by sonar. For four days an unidentified submerged object played a cat-and-mouse like game with the Naval task force. The object was tracked by sonar travelling at speeds up to 170 miles per hour and manoeuvring to depths as great as 27,000 feet.

The (probably non-existent) triangle has fuelled the careers of writer Charles Berlitz, and singer Barry Manilow, and launched a dozen third rate books which claim to be a compendium of the world's most mysterious mysteries. This present volume isn't one of them.

However, the thing that has interested me most, has been the successive wave of strange animal sightings across the island.

Strange animals are a well known part of the island's culture. *Vejigantes - papier mache* masks worn during the carnival, are the island's most popular craft. Tangles of menacing horns, fang-toothed leering expression, and bulging eyes of these half-demon, half-animal creations makes these masks frightening, particularly at night-time. During carnival time, masked revellers roam the streets in their bat-winged costumes either individually or in groups.

Ponce is the major centre of mask making on the island. The masks are traditionally black, red and yellow in colour, all symbols of hellfire and damnation. Nowadays pastels and other colours are also used. The *vejigante* sport at least two or three horns; however some of them are decorated with hundreds of small horns of all shapes and sizes. Masks can be simple or elaborate, and also depend on the character represented. *El Caballero* (the knight), *los vejigantes*, (the devils) *los viejos* (the elders) and *las locas* (the crazy women) are the four main costumed characters.

The origins of the mask, and the carnival, date back to medieval Spain and Africa. In the 17th Century a series of processional parades in Spain were intended to terrify sinners with marching demons so that the people may return to church. Originally the *vejigante* were demons that appeared at the festival for the patron saint, Santiago. In Puerto Rico the traditional of the demon parade was combined with the masked ceremonies of Africa as figures of resistance against colonialism and imperialism.

However, this diabolical bestiary is not confined to the innermost recesses of the collective imagination of the islanders. For at least the past fifty years there have been a whole series of weird animals, many of which make no zoological sense, reported in the *El Yunque* region of the island.

In the 1950s a small bigfoot-type creature (if this is not an oxymoron) was reported on a number of occasions in the mountain rainforest. Ten years after, weird hairy dwarfs were reported. This was at a time when such things were being reported all across Central and South America and, unlike their precursors, were often seen in conjunction with UFO activities. A typical mainland South American incident took place in Venezuela in 1954, near the capital Caracas. Two men claimed to have wrestled with a hairy creature that was seen to emerge from a luminous craft. One of the men, Gustavo Gonzalez, claimed to have been thrown fifteen feet through the air by a tiny dwarf like creature, who despite its size possessed incredible strength.

One particularly striking account comes from Barrio Rabanal in 1979:

This and next page: a range of *Vejigantes* masks

Island of Paradise

"Three juveniles playing behind their house by the roadway suddenly heard a noise coming from the brushy area next to the road. Then a bizarre creature jumped from out of the brush and into the middle of the street. The being was described as being three-foot tall with its body covered with brown hair. It had two huge strong legs, resembling those of a rabbit, it lacked any arms. It had a relative small head with two huge staring eyes and a large mouth with two protruding fangs. The creature had several fleshy tube-like protrusions on its body, mostly on the chest area from which a green liquid like substance was oozing out. The creature made some strange noises and began salivating heavily; it then ran into the brush again and disappeared towards a nearby river."

There are dozens of such incidences on record, and this is neither the time nor the place to go into them. However, as the monster sightings increased in frequency on the Island of Paradise, so did UFO activity, and as will be seen in a later chapter, there does indeed seem to be some correlation between the two.

In the early 1970s, *aves d'diablo* (devil birds) were reported. Interestingly, these seemed to have more to do with some of the strange reptilian 'pterodactyl' sightings from the Rio Grande valley on the Texas/Mexico border in the late 1970s, than with anything else known from the Caribbean. They were described as being like big, black, turkey vultures with leathery wings, which had somehow had the metal spurs used by professional cock-fighters grafted on to their beaks. Whilst many species of bird have an egg-tooth on their beak when they first hatch (to facilitate breaking through the natal shell), and hornbills, toucans and some species of pigeon (most notably the long-extinct dodo) have bizarrely ornamented bills, to my knowledge no species of bird that has ever existed has a true horn. [1]

A few years later the devil birds had flown off into the skies above Charles Fort's celestial Sargasso Sea, never to be seen again. In the mid-1970s they were replaced by a new monster - the notorious *Vampiro de Mocha*; an entity whose activities began in the town of Moca's Barrio Rocha, where it killed a number of animals in a grisly fashion never seen before. Fifteen cows, three goats, two geese and a pig were found dead with strange puncture marks on their hides, indicating that some sharp object - natural or artificial - had been inserted into the hapless bovines. Autopsy reports invariably showed that not a single drop of blood remained within the animals, as if it had been consumed by some predator. Police officers were adamant about ascribing the deaths to dogs, since they correctly believed that not even the wildest of feral dogs could climb some of the fences surrounding the dead animals' pens.

On March 7, 1975, a cow belonging to Rey Jiménez was found dead in Moca's Barrio Cruz, presenting deep, penetrating wounds on its skull and a number of scratches around the wounds on its body. Jiménez's cow was added to the growing list of victims, which now totalled well over thirty.

As the number of victims grew exponentially, the Moca Vampire acquired an identity of its own, much in the same way that the chupacabra would twenty years later. Speculation as to its nature was rife: many believed it was a supernatural "bird", like the one seen by María Acevedo, a Moca resident who noticed that a strange animal had landed on her home's zinc rooftop in the middle of the night. According to Acevedo's testimony, the bird pecked at the rusty rooftop and at the windows before taking flight, issuing a terrifying scream.

There was then a wave of unexplained attacks of livestock throughout the island.

---

1. These monster reports are from three main sources: http://www.thelosthaven.co.uk/1989Cont.htm (a website called 'The Humanoid database'), Charles Bowens brilliant *The Humanoids* (NEL, London, 1969), and the aforementioned books by Corrales. The *Vampiro de Mocha information is mainly gleaned from an article in the 1998 CFZ Yearbook*. However, I have been collecting bits and bobs on the fortean aspects of Puerto Rico for many years, and some of the stuff in my files is unreferenced, but as - once again - this stuff is just really to fill in the back-story to the main events of the book, the fact that some of these accounts are not properly referenced doesn't really matter.

Salvador Freixedo investigated these at the time and wrote [1]:

"During an evening in which UFOs were sighted over the town of Moca, two ducks, three goats, a pair of geese, and a large hog were found slain the following morning on a small farm. The owner was going insane, wondering who in the world could have visited this ruin upon him. The animals betrayed the wounds that have become typical of this kind of attack, and of course, they were all done with incredible precision. I did not doubt for one moment who could have been responsible for the crime... I got in my car and visited the area immediately, and realized what was filling the animals' owner with wonder and fear: there wasn't a trace of blood in any of the animals, and in spite of the fact that the dead geese had snow-white feathers, upon which the slightest speck of blood would have shown up immediately.

"Over the next few days, the newspapers continued reporting on the growing number of dead animals found in the region. No explanation could be found for these mysterious deaths. I visited the rural areas on various occasions to investigate the events firsthand and found that the farmers were as intrigued by their animals' deaths as they were by the enigmatic lights they could see in the nocturnal skies. One of them told me that the lights reminded him of the revolving lights on top of a police cruiser.

"During one of my forays, I was able to see a black and white cow spread out in the middle of the field. I got out of the car and tried to reach the cow, which wasn't easy. The dead beast had characteristic wounds on its neck and on its head. Skin had been pulled back on one side of its head, as if by using a scalpel, and the opening to one of its nasal orifices was missing, although there was no indication of rending. In spite of the whiteness of its head, there wasn't a single drop of blood to be seen. The farmer who escorted me could not stop wondering what had caused his cow's death. He related how that very same night he had heard his dogs barking furiously, and that a blind elderly woman who lived on the edge of the field had told him that the cattle, which ordinarily spends the night outdoors, had kept her from getting a good night's sleep due to their frantic, maddened running from one end of the field to another."

There were also sightings of a green hairless dwarf, and as farmers are wont to do, they put two and two together, made the number of the beast, and assumed that the former were the product of the latter. The attacks only continued for a period of months, and in the 1980s yet another monster appeared on the scene.

The plantain farmers of the high grassland plateau surrounding Canóvanas reported a strange semi bipedal creature covered in spines, which was supposedly attacking their crops. This is probably the most ill-documented of the Puerto Rican pantheon of monsters.

I only became aware of this during my first visit to the island when Graham (who was already horribly hungover) and I filmed a riotously drunken interview with two American journalists in a bar in one of the seediest parts of old San Juan.

They were the authors of a Spanish language booklet on the chupacabra, and whilst they had been minded to treat the whole affair with levity, they gave us some interesting snippets of information, which were - eventually - to help us come to the hypothesis described in this book.

---

1. Like many of the UFO and mutilation reports in this chapter this is Nick Redfern pers. Comm referenced: http://www.ufomystic.com/the-redfern-files/moca-vampire4/

Island of Paradise

The island media soon christened the beast "Comecogollos" - the Banana-Tree Eater [1].

The silly name did nothing to assuage the monster's temper. Whereas some people described a spined animal, other eyewitnesses described it as a manlike, hairy creature weighing some sixty pounds, strong enough to kill a dog and a goat, and tear its way through plantain groves. Manuel Rivera, a planter and businessman in Lagunas, complained to the press that not a single government agency had paid attention to the matter, and that the police refused to respond to calls involving the strange creature. A number of goats slain by the hairy being had to be buried when no official agency turned up to perform autopsies.

In the early 1990s there were a series of singular sightings, which were reported and investigated by that doyen of Hispanic high strangeness - Scott Corrales. According to him, in Montana Santa during 1992:

"At the same time of the miraculous apparition of the Blessed Virgin in this hilltop, a number of unidentified artefacts were reported in the hill's vicinity and even captured on film. Other witnesses reported brilliant, disk-like objects flying overhead. One of the witnesses, Delia Flores, reported that she and other worshippers were surprised to see a beige van parked in the area of the religious sanctuary on the hill. Its occupants wore orange fatigues with NASA insignia, and the vehicle's Spanish speaking driver told them that the van contained a most unusual cargo - a simian creature captured in the Caribe State Forest (El Yunque). According to Flores, she and the others saw a covered cage that contained something "struggling to get out." The driver added that the creature was being taken to a secret primate research laboratory located somewhere in the island, where investigation on this sort of being was being conducted. A local farmer discovered one morning that a number of plantain plants on his property had been destroyed by an unknown creature that left a number of deep footprints, attesting to its massive size and weight. Other residents indicated that they had seen a "hairy figure" run away from the area in the darkness, but they could not described it in detail."

A similar report took place in Maracal the same year:

"Gerardo Rosario was weeding his property when he suddenly heard a noise to the side, upon investigating it; he noticed a hairy creature climbing up the hillside. It was about 5-feet tall, hairy and was accompanied by a smaller hairy creature just like it. He could not make out their faces because they were climbing sideways, but he noticed hair covering their features except around the eyes and cheeks. Another witness, a 12-year old boy on his way to school was distracted by some odd sounds coming from the roadside. When he went over to investigate he discovered two creatures sitting on a large boulder in a mountain stream at the bottom of the ravine. The larger creature walked around the top of the boulder, as if keeping watch, while the smaller figure remained seated. The larger creature emitted the moan that had drawn his attention in the first place, causing the witness to become frightened and flee the area."

And again in Lomas Verdes...

"One night while worshippers fervently prayed in the heavily wooded area, a five-foot tall, muscular figure covered with brownish hair raced between the trees in a zigzag pattern. Also in the same area, while religious apparitions were in full swing, sightings of large headed grays were also reported"

---

1. *Mysterious Creatures - a guide to Cryptozoology* by George Eberhardt (ABC Clio, 2002)

There were also reports of a pterodactyl-type creature, such as this one from Olivares:

"Heriberto Acosta watched a huge winged creature resembling a pterodactyl or a gargoyle descend from the sky and perch on top of a nearby tin roof. The grotesque creature had fleshy wings with feathers and what appeared to be scales. It had a long neck and a beak filled with sharp teeth. It eventually flew away."

The next anomalous animal of this most peculiar year was a "Merman". Although most zoologists would be perfectly happy to consign such things to the realm of storybooks, and whilst most cryptozoologists would feel happier doing so, these stories surface with a worrying frequency. As I write this there are reports from the Caspian Sea of an "amphibious human", which sound strangely similar to this Puerto Rican report from Playa Media Luna, Vieques Island, in 1992:

"Mr. Anibal Perez and his young nephew had come to the beach with their grandmother and were enjoying themselves when suddenly and apparently out of the depths of the waters a "young man" about 12 years of age, emerged and approached the two witnesses. The stranger appeared curious and friendly and followed the two witnesses apparently looking for companionship. They asked him who he was but the stranger only emitted a sound similar to that of a "dolphin." Suddenly the boy dived into the waters and seemed to swim down to the depths. Moments later he re-appeared again bringing some sand in his hand, which he offered to the two, now befuddled witnesses. Apparently he wanted the men to "eat" the sand, but they demurred. They were then even more surprised to see the "boy" eat the sand like material. The stranger was described as completely human-like, somewhat thin with long black hair, wearing light colored yellow shorts. As the two witnesses swam about, the stranger seemed to imitate their every move. He dived into the deep again, this time returning with some algae, which he also offered to the witnesses. Emitting strange whistling sounds he seemed to want to communicate with the witnesses. After some questioning by the witnesses, the stranger began to imitate their speech, repeating everything that was said to him. After a while he seemed to be able to speak some coherent words and said he was from "the bottom of the sea." A few minutes later he stared at the two witnesses then bid them farewell saying "Goodbye humans," he then dove into the waters never to be seen again."

To cap off the high-strangeness of 1991, government agencies on the offshore island of Culebra, to the west of Puerto Rico, found themselves faced with the appearance of a "mystery cat" - a one hundred and fifty pound beast, grey in colour - seen by personnel of the Natural Resources Department and of the Conservation and Development Agency on Culebra's Playa Flamingo.

In the mid 1990s the most notorious of Puerto Rican monsters than reared its ugly head. Known as the chupacabra (literally: goatsucker) it linked yet another outbreak of animal killings with a spate of sightings of a bipedal, spiny creature very similar to that which had attached banana plantations a decade before. Once again, the epicentre of these sightings was Canóvanas, but the sightings took place all along the Canóvanas River, which rose from a tiny spring high up in the mountains and continued its inextricable progress to the Atlantic Ocean. At the time of our first visit to the island, Graham and I did a fortnightly radio show for *BBC Radio Devon*, and we did two outside broadcasts from our travels. During one of these I described the chupacabra as being akin to the computer game character Sonic the Hedgehog on acid, and mildly to my embarrassment, the description has stuck.

Scott Corrales describes the first known report:

"Madelyne Tolentino and her husband, José Miguel Agosto, have the distinct privilege of beign the first witnesses to the creature. During the second week of August 1995, at approximately four o' clock in the afternoon on a weekday, Ms. Tolentino looked out a window at her home and saw a young man walking backward with an expression of indescribable fear on his face, as if something horrible were about to pounce on him. She then noticed that a strange creature was approaching the house at a moderate pace, allowing her to take a good, long look at the aberration. Whatever it was stood four feet tall and had a pelt covered in a mixture of colours ranging from brown to black and ashen grey, as if it had been burned. Tolentino added that the entity had gelatinous dark-grey eyes and spindly arms ending in claws. "To me, it couldn't be anything from this world," she would later tell reporters.

Her momentary fascination with the entity came to an end when she realized the enormity of the experience. Shouting, she called for her mother to witness the surreal event. Her mother would later add that it had a coppery plumage running down the length of its back and that it moved in a series of short hops, like a kangaroo, but lacking the marsupial's characteristic tail. The creature ran into an overgrown field, and Ms. Tolentino's husband and other residents of the same street gave chase, but the creature was nowhere to be found."

The descriptions varied, and became more outlandish as a peculiar game of Chinese whispers was played across the island. The animal was described as being able to fly, being able to apparate and disapparate at will, and alternately having the face of a guinea pig or the visage of Satan Mekakreig Himself. Then something extremely interesting from the Socio-Political viewpoint at least – happened. For the first time reports of what had been purely a Puerto Rican monster started being echoed on the mainland of Central and South America. First in Mexico, then in Brazil, then Chile, then the Spanish speaking parts of Florida, and by the turn of the century wherever there was a strong Latino culture in the New World there were chupacabra reports. Something else that is notable, however, that is the further away these reports took place from the motherlode in Canóvanas, the more disparate the sightings and reports became. UFO magazines across the globe produced photographs of a pterodactyl-like creature allegedly photographed in Chile and a strange gargoyle allegedly photographed in Argentina. I cannot prove it, but I am convinced that the pterodactyl came courtesy of those jolly nice people at Adobe Photoshop, and I *know* that the oft-circulated gargoyle photograph was actually of a carving found on a wall in Wookey Hole caves in Somerset. Further north in Arizona and the neighbouring states a series of chupacabra carcasses turned up. One was an olive baboon, which had obviously escaped from a private collection, and another was the gas-bloated and horribly putrefied corpse of a chow dog. As the years progressed it became obvious that the term chupacabra had become a catchall term across Hispanic America to cover any monster or bogeyman. This was nothing new in the annals of Forteana. The Australian word 'Bunyip' had originally been used to describe long-necked lake monsters in central Victoria. By the end of the 20th Century bunyip too had become a catch all term for anything weird and had even been used to describe an alien 'grey' in a mid '90s episode of *Home and Away*.

Occasionally one of these spurious chupacabra reports appeared to be of a genuine cryptozoological creature. In 2003 there were a series of reports from Argentina of a weird bipedal creature the size of a man and covered in hair. A newspaper report of the time from *El Tribuno Digital and Inexplicata* 4/16/2003 read:

"New testimony regarding the presence of a strange hairy biped, sporting long claws and an unusual ferocity, were added to the long list of accounts that the police have gathered in" Rosario de la Frontera, a small city in northern Argentina "after the judge of the Court of Instruction, Mario Dilascio, mandated a final investigation into the subject last month."

Island of Paradise

"The first reference to the case was provided by a couple in a forested area, Arroyo Salado, two kilometers (1.2 miles) to the west" of Rosario de la Frontera "and which serves as a 'Lovers' Lane.' They claim to have been attacked by a strange entity standing two meters (6 feet, 6 inches) tall, covered in hair, with razor-sharp claws and 'bare buttocks.'"

"But now there is more: a family surnamed Pereyra told authorities that whenever going to the municipal dump in their pickup truck, 'a beast' described as a 'large monkey' crossed their path, climbed over the vehicle's hood, and then lost itself in the thicket, leaving claw marks on the (truck's) chassis."

In another case, "businessman Raul Torres and teacher Hugo Rodas, who teaches in San Antonio de Cobres, told the chief of the volunteer fire brigade, Jose Exequiel Alvarez, that they found something very unusual at El Duraznito."

"A large unknown animal lay by the roadside, apparently run over by a vehicle. They got out of their car to take a look, turned it around using a stick, and were astonished. They had never seen anything like it. It had amazing claws, was like a human, measured 1.5 meters (4 feet)" in length, "had a bearlike snout with enormous fangs and genitals identical to those of a male human."

"Alvarez has led several expeditions to find the animal and claims having seen it through binoculars. 'What I saw resembled a gorilla very closely but,' he asked himself, 'what would an African anthropoid be doing here?'"

"The volunteer fireman added that 'we will be visiting El Duraznito to see if we can find the remains (of the creature--S.C.) reported by those people.'"

"Furthermore, a woman surnamed Galvan claimed having seen it in Los Banos, behind the mountains surrounding the thermal springs."

"'I don't think we're dealing with a single specimen, but with a family of them,'" Alvarez said.

The descriptions of its sickle-like claws led cryptozoologists at the Centre for Fortean Zoology to excitedly add it to the slowly growing body of evidence to support the idea that prehistoric giant ground sloths had escaped the extinction trap and had survived until the present day.

In my first book about the chupacabra, I reached the conclusion that it was not a flesh and blood animal. I hate to use the words 'paranormal' or 'supernatural' in this context. They are both massively overused and misunderstood terms. I am a scientist and would like to stress that I don't believe in hocus pocus or mumbo jumbo. However - as the person who first coined the term 'zooform phenomena' to cover things which *look* like animals, but which cannot be explained within accepted zoological frames of reference - I feel that I must defend myself by saying that I believe such things are governed by natural laws of physics such as those which govern every conventional life form on the planet. It is just that at the moment we don't know what those laws are.

But as I sat with Nick Redfern that hot July night in San Juan quaffing margaritas, smoking cigarettes (which I was supposed to have given up), and generally talking about nothing in particular, I was still fairly convinced that there could be no mainstream zoological explanation for the chupacabra of Puerto Rico. It only took a few days for me to realise that I was completely wrong.

My first trip to the island had been financed by British Channel 4 TV, who wanted an entertaining bit of pop science for their series *To the Ends of the Earth*. Sadly this ain't what they got. Although I can do pop science with the best of 'em, from day one there was a serious conflict of interest between the director and me. He seemed more intent of making a comedy programme of a fat bloke (me) getting in and out of smaller cars, and was not interested in following our real investigation. This was best illustrated when we visited a farm whose sole reason for existence was to breed roosters for cock-fighting. Apart from the fact that I abhor bloodsports and flatly refused to attend a cock-fight, I also seriously dislike the sort of Englishman abroad who insists on treating the globe as if it was a leafy corner of the Home Counties and imposes his own personal cultural morality upon the people he meets.

Johnny Foreigner may indeed be a Rum Cove, but I for one have no right to tell him how to live his life, and this rather seedy establishment had been the location for a major chupacabra attack only a couple of days before our visit. Whereas the film crew had filmed me interviewing the proprietor and his mate and wandering around the farmyard looking intrepid, they had refused to film Graham and me retrieving the one chicken carcass, which remained, and carrying out a makeshift autopsy. There is - quite possibly - a code of conduct for British broadcasters, but whilst I am quite prepared to admit that the sight of a fat bloke stripped to the waist and sweating like a pig whilst trying to dismember the exsanguinated cadaver of an unfortunate chicken may not be quite within the bounds of accepted taste for mid-evening broadcasting, the fact remains that they ignored our discovery and instead filmed a piece of semi spooky nonsense for the opening credits which featured the farmer and his mate shot with spooky backlighting, leaving Graham and me to discover the most important piece of evidence which was garnered from our first trip to the island. It was at that point that I gave up on trying to make any sort of credible documentary.

Seven years on there is no point in recrimination. I feel that the inclusion of this sequence would have made for a much better film, but luckily, as it turned out, Graham and I filmed the whole thing for ourselves and even took samples of the blood soaked feathers back to England. At the time, one of the most popular shows on British television was *Buffy the Vampire Slayer* and for several years I dined out (and got a cheap laugh) with the line that whereas I would have much rather seen Sarah Michelle Geller stripped to the waist performing a vampire autopsy I, for one, was probably the only real vampire hunter that anyone was likely to meet.

The chicken was found where it had been killed - on a dry dusty surface beneath a wizened cherry bush. Puerto Rican cherries are one of the most delicious fruits I have ever eaten. They taste as if someone had taken one of the jars of maraschino cherries that your father-in-law always buys at Christmas, and shoved in a generous dollop of molasses and a tot of Irish whisky. However, call me old-fashioned if you like, but even I couldn't face plucking and eating these delectable fruit from a bush under which only 48 hours or so previously a putative vampire had allegedly at least, done his funky thang (or should that read *fang*?). Graham - with much protest - crawled on his belly like a disgruntled snake, and filmed the chicken carcass *in situ*. There was no sign of blood on the ground. Indeed, when we took the chicken out and began to cut it up, we found no signs of blood at all.

Over the years, as I have carried out my sordid career I have conducted autopsies on everything from a woodlouse to a dolphin. During my career as a nurse I laid out corpses on several occasions. I have been present at open heart surgery, and a variety of less exciting surgical procedures and I doubt whether even my mother - God rest her soul - would have accused me of being squeamish. However, on this occasion I have to admit that faced - for the first time in my life - with something whose death, I believed then, could have been laid at the door of some supernatural entity, I felt the hairs on the back of my neck stand on end. In many ways it was easier for me that the film crew were fannying about doing something else, for once on the trip I didn't have to play for the cameras. I examined the body in some detail. Again, during my long and chequered career I have examined dead animals in various states of decay. As a child in Hong Kong I had seen how the ravages of a tropical climate, and the attacks of micropredators could very quickly make a freshly killed animal unrecognisable. I was aghast. Apart from the damage caused by feather mites (and upon examining the few living chickens which remained, I could see that feather mites were endemic on the farm), there was no sign of micropredation at all. The only wounds on the creature were a hole - roughly the size of a biro - in the breastbone and two puncture marks which appeared to have been caused by fangs on the neck. The creature was entirely drained of blood. Most spectacular, when I opened the thoracic cavity, and later the body cavity itself, I found that the membrane between the two was unbroken, but the liver appeared to have been removed. The liver is the most highly vascular organ in the body, and is full of blood. Whatever had taken the creature's blood had also removed this organ, and whilst I could guess why, I was damned if I knew how!

At the time I think it is fair to say that I felt that this was the most momentous discovery of my career to date. Together with the macabre events surrounding some bloodless goats in a small village in the Puebla desert in Mexico, this event became the cornerstone of much of my public speaking and theorising for the next few years. I was convinced that I was party to an event that had astounding implications for mankind. It wasn't until the autumn of 2003 when I was sitting on top of a red sandstone mesa in the Nevada desert with a suspiciously long cigarette in one hand that I realised that I might have been jumping completely to the wrong conclusions. I had been at a UFO conference in Las Vegas - more of which later - and had, as is my habit, managed to sneak away with a couple of pals to see the local scenery and to get away from the miasma of pernicious bullshit which usually surrounds such events. Together with Nick Redfern, and a gaggle of the younger and more *avant-garde* conference goers who followed us, I drove into the heart of the Valley of Fire National Park. For about a quarter of an hour we all did our own thing and, as I always do, I wandered off to look for the local wildlife. Turning over stones to look for small lizards and scorpions I came across an ants' nest. Such things have always fascinated me. Ants, as the late, great T.H. White pointed out in his classic children's book, *The Sword in the Stone* - have a very complex, almost machine like, society, and as one of life's natural anarchists I derive a great deal of pleasure from watching such an organised hegemony in action. There was a column of worker ants travelling earnestly to and from the nest. Following them I found, only a few feet away, a dead bird, or what remained of it. It was desiccated from the sun, and although I was unable to identify it, it was a quail or small partridge. To my amazement I saw that whilst the ants were obviously feeding off the unfortunate creature's carcass and carrying bits of it back to the nest, they were not leaving any marks of micropredation upon the skin or feathers. Instead, they were eating the animal from the inside out, gaining entrance and egress through the cloaca.

Six months later, whilst sitting drinking a cup of tea in a back garden in Illinois, I saw exactly the same mechanism at work - this time on the body of an American robin. Could I have discovered a way that the liver could have been removed without any obvious exit wounds? Borrowing one of my hostess' kitchen knives I cut the bird open and found - not altogether to my surprise - that the liver of this creature was also missing.

Although whatever had killed the chickens in Puerto Rico was certainly not ants, suddenly the missing liver did not assume such momentous importance as I had claimed for the preceding six years.

During the latter part of the last decade of the 20th Century, one of the most popular television programmes and one, which collaterally made me a lot of money, was the *X-Files*. It told the story of two American FBI agents who travelled around solving mysteries and it introduced a whole new generation to the wonders of the fortean omniverse. Large proportions of its content were complete bunkum and for my tastes it was too obsessed with UFOs, Government cover-ups and conspiracy theories but it did make its mark on popular culture, not the least with its weekly watchword that "the truth is out there".

I have always taken this to be one of the slogans of the CFZ. The truth is indeed always out there, but in my experience; it is usually not what people expect it to be. I have always been a great admirer of song-

The author in the Valley of Fire National Park, Utah, November 2003

Valley of Fire National Park, November 2003.
L-R: Matthew Williams, Kenn Thomas, Greg Bishop, the author, Nick Redfern

writers and authors for whom their life and their art are interchangeable. Writers like John Lennon and Phil Ochs made contemporary records in what Ochs calls a newspaper headlines style. Quite often the results were pretty terrible, as in John Lennon's three albums of anecdotal "unfinished" music, but they were honest and they were brave. I have always tried to do my writings with similar integrity.

There are writers within the fortean universe who, when they decide that a theory that they have been working on is wrong, and has now been superseded by a new and better one, proceed to rewrite their earlier books in an almost Stalinist manner. I will not do that. My first book on the chupacabra is, if nothing else, a document of how I was, what I thought and what I felt at the time. I have no intention of going back and rewriting it. The fact that this book almost entirely contradicts the Puerto Rico chapters (at least!) of its predecessor is just the way that the fortean cookie crumbles. The fact that my whole theory on the nature of the chupacabra in Puerto Rico stood and fell on the flawed evidence of one desiccated chicken has not prompted me to rewrite the past.

However, my discovery on that red sandstone mesa in Nevada on that late November evening, was one of the pivotal moments towards putting together the theory that is the basis of this book.

I may have solved the mystery of what had happened - or to be more exact what *could* have happened to the missing organ, but I had not solved the mystery of what had killed the animals in the first place. I had also got absolutely nowhere in my efforts to explain what the spiny bipedal creature that I had dubbed 'Sonic the Hedgehog on acid' actually was.

Everywhere I went I was confronted by eyewitness accounts such as this one from 1997:

"Mysterious attacks on domestic farm animals around the town of Utuado, Puerto Rico, 40 miles (64 kilometres) southwest of San Juan, have triggered rumours of renewed predation by the Chupacabra or legendary "goat sucker." Forty-two large white rabbits, some chickens and a duck were found dead on a farm in Utuado on Thursday, November 20, 1997. The dead animals had twin perforations mostly in the stomach region and on their feet. Most of the perforations were triangular. According to researcher Scott Corrales, "One rabbit had its stomach split, an incision so precise as to only have been made by a surgical instrument or by an expert surgeon. No trace of blood remained in any of the dead animals."

UFO ROUNDUP, Volume 2, Number 46 November 30, 1997

My first stay on the island had convinced me that the chupacabra was a very real phenomenon. Hopefully, my second trip would bring me closer to identifying it.

On the whole I enjoyed my first journey to Puerto Rico. I didn't enjoy the filming process, and was not particularly proud of the resulting documentary. The book I wrote about the trip has its moments, but is basically a journal of how a younger and more foolhardy Jon Downes stumbled around the region jumping to conclusions and doing things that the older and wiser Jonathan would probably never dream of.

The one enduring legacy of the 1998 sojourn in the Island of Paradise left was that I fell in love with the island. It reminded me of my childhood in Hong Kong - a place and a time when I was probably more nearly happy than I have been at any place since. There was no great difference from the foothills of El Yunque and the mountainsides of Victoria Peak where I played as a child. Visiting the island, albeit under less than ideal circumstances, had for a few brief days allowed me to revisit my childhood. I have spoken to other people who have visited Puerto Rico and found that my experience is far from being unique. It is a glorious land, a Lilliputian kingdom that, although allied politically and economically both to the United States and to the independent countries of the Caribbean, somehow stands apart from them all in its own surrealchemical landscape. In J.M. Barrie's *Peter Pan* he says that the country of Neverland, which is where people go in their dreams, is different for each of us who visits. Some children, he says, dream of pirates in a great lagoon, others of flamingos and others still of Red Indians. Since 1998, when I close my eyes, I find myself more often than not transported to the Island of Paradise.

The Margarita Diary,
Chapter 1 Comments:

Since Jon so skilfully captures in this chapter the essence of what makes Puerto Rico so magical - in terms of its history, its culture, its people, and not to forget its overwhelming weirdness - I will refrain from commenting specifically on such matters myself.

Instead, I will say this: it is these very ingredients, namely the far-off locations with exotic names, the vastly different cultures, and the mysterious locales that collectively led me to know that this was an expedition, an adventure, and an experience that I could not afford to turn down. Truly, Puerto Rico is a magical locale that attracts the adventurer and the thrill-seeker like no other. And given that it was a veritable hotbed of activity of the ufological, vampiric and downright uncanny kind, what else could I, or indeed we, do but welcome the weirdness with wide-open arms.

If Jon and I were going to spend a week hunting vampires courtesy of the Sci-Fi Channel, then there was no better place to do it than deep within the heart of the island of paradise, and while regularly fuelled by the finest of local cuisine and those ever-present chilled margaritas. Onward! **NR**

Chapter Two
Return of the exile

A lot had happened - both to me personally, and to the Centre of Fortean Zoology - between 1998 when I first went to Puerto Rico, and 2004 when I returned, an older and wiser man. Whereas in 1998 the CFZ was just a bold concept; we published a quarterly magazine, had about 100 members and did little bits of research in the UK, by 2004 we were the largest and fastest growing cryptozoological organisation in the world. We had taken part in expeditions across the globe, and that year alone I had been looking for bigfoot in the Texas swamps, and chasing mystery cats in the American Midwest, while Richard was leading his second expedition to Sumatra in search of the elusive *orang pendek* - an unknown species of upright walking ape. At the 2003 *Fortean Times* Unconvention alone, we had taken over 60 new members and over £2,000 in donations. By anybody's terms we were now a force to be reckoned with. On a personal level my life had changed as well. I have written elsewhere of my health problems and have no intention of repeating myself here, but by the time I left England in July 2004 I could only walk with a stick, and my hand baggage contained an impressive battery of medication. During my first trip to the island I was regrettably bullish in my attitude, and Graham and I spent much of our sojourn in Central America trying to party like it was 1999 when it was only 1998. Now I was older, and wiser, and determined not to repeat the mistakes of my first trip.

The zoological mysteries of Puerto Rico had been of enduring fascination to me, and in the intervening years I had spent of much of my time of gathering information and opening voluminous files on the subject of the island, its history, ecology, and wildlife. I was invited to attend a UFO conference in Las Vegas in November 2003. I used it as an excuse to talk about some tentative experiences which are explored more fully in this book, and despite the fact that the audience - to a man - were steadfastly of the "I want to believe" brigade, and who were adamant in their belief that unidentified flying objects could only be wayward emissaries from outside our solar system, my talk went down reasonably well. As is always the case at these affairs, the most interesting discussions take place at the bar, not as a result of the on-stage question and answer session, and during my sojourn in "sin city" I was lucky enough to meet two people whose testimony filled in big gaps in my burgeoning theories.

I ended my talk - as always - with an appeal for funds. I badly wanted to return to the island and continue my research, but although the CFZ was relatively solvent by that stage, financing an expensive trip to Puerto Rico on top of our other commitments for the year was out of the question. About half an hour after I had finished my talk I was in the bar with Peter Robbins (co-author of the classic Rendlesham Forest book *Left at East Gate*) and Kenn Thomas (editor of the conspiracy journal *Steamshovel Press*). As is usually the case at these events, our conversation was about anything and everything apart from UFOs. As far as I remember we were quaffing margaritas (provided by a benevolent management at 99c a glass with the aim of getting the punters as pissed as they could so they could lose all their money at the crap tables) and talking about - of all things - Timothy Leary's life as a political exile, and the Apple records solo albums of Yoko Ono. We were approached by a short, podgy middle-aged man of obvious oriental descent. In all the years that I have been travelling around the world I do not think I have met anyone with such a broad grin. He looked like a cross between Pacman and an Argentine horned frog, and he smelt strongly of garlic. He also had a cleft palate, which made conversation with him difficult, but it turned out that he was a Japanese-American realtor who, during his student days, had spent some time working as a volunteer helper in El Yunque National Park. During my talk I had listed the pantheon of Puerto Rican monster sightings that I cover in some depth in Chapter One. Much to my great delight, my new friend, whose name it turned out was Benny, had actually witnessed a pair of the *Aves d' Diablo* during the late 1970s.

"They were really strange, man", he said in a tremulous voice. The trouble was that as he got more and more excited, his speech impediment became worse and with every word - especially when he was speaking in Spanish - he showered us with a revolting mixture of spittle and half-masticated nachos. *"We were working a forestry plantation in the hills above Aguas Buenas"......*

My ears pricked up. Aguas Buenas is an area particularly dear to me, and the location for one of the particularly vexing zoological mysteries that I hoped to solve. *"It was really strange because it was just after dawn and all the birds in the forest had been - how do you say it? - screeching. Then suddenly it all fell quiet and I could a strange booming sound."*

I sat bolt upright. The same thing had happened to me - twice - in the Northumbrian wood only the previous January [1]. The dawn chorus, and later the dusk chorus fell silent as if God Himself had pressed the mute button on his TV remote control, and I too had heard the strange booming noises. Within minutes I had seen a monster. It looked as if Benny - back in 1979 - had experienced a frighteningly similar event.

"Then I saw them ....." he spluttered, causing such gastronomic devastation in his wake to make my two companions withdraw from the scene of battle. In the best tradition of tabloid journalists from the News of the Screws they made their excuses and left. *"They were two enormous birds. They looked like turkey vultures but I swear to you that they were the size of a microlight. They were jet black, they had red glowing eyes, and they had strange horns on their beaks a bit like a rhinoceros. But that wasn't the scary thing........"*

He took a deep breath, reached into his jacket pocket and got out a pouch of rolling tobacco. Slowly and laboriously, with the manual dexterity of an anteater on Quaaludes [2], he began to roll himself a cigarette. It was obvious that he was one of these people who have a 110% concentration to the task at hand, and whilst I wanted to slap the tobacco out of his hand and tell him to "get on with it, for God's sake" I

---

1. Downes, J *Monster Hunter* (CFZ Press, 2004); Arnold, N *Monster!: The A-Z of Zooform Phenomena* (CFZ Press, 2007)
2. QV: Methaqualone, mandrax

couldn't, so I lit a cigarette of my own, took a deep swig of my margarita, looked reproachfully over at Peter and Kenn who, by this time, had joined up with Nick Redfern and a gaggle of scantily clad Las Vegas showgirl types at a table at the other end of the bar, and looked as if they were having the time of their lives.

Finally, Benny finished rolling his cigarette. He put it in his mouth and asked me for a light. He reached over to borrow my cigarette lighter and knocked the pitcher of margaritas on to the floor with a resounding crash. Suddenly we were surrounded by the serried ranks of hotel domestics who - disregarding the fact that the whole affair was as a result of Benny's clumsiness - insisted on behaving as if the whole episode had been their fault. It took a full ten minutes before - having been provided with a new table, a fresh pitcher of margaritas, a clean ashtray, and a new bowl of nachos for Benny to graze upon - we were ready to continue with the saga of the devil birds of Aguas Buenas.

Because of the delay, I had to talk Benny through his experience from the beginning, and - to tell no word of a lie - just as he was about to tell me what "the really scary thing" was, then he reached in his pocket for his tobacco and started making moves to roll another cigarette. "For crying out loud", I muttered under my breath, and pressed one of my precious UK Benson and Hedges upon him. He peered at it with distrust as if the very act of accepting a foreign made cigarette was going to bring down the thought-police of George W. Bush's regime upon him. Manfully resisting the urge to slap him about the head and neck, I lit a cigarette for him and urged him to continue his story…..

"The really scary thing was, that when the sound went, time seemed to slow down. I could see these creatures flying above me and above the jungle, but they were flying in slow motion. They didn't look like real birds. They were like badly animated puppets…. I don't know if you remember the Gerry Anderson TV series Fireball XL5, *but there was an episode of that with giant birds at the edge of space and although the were flapping up and down on strings, when I was 7 it was the scariest thing I had ever seen in my life. When I was 19 in the hills above Aguas Buenas it was even scarier."*

Despite the fact that he was most obviously a klutz of the first order, there was no denying Benny's sincerity. We talked for a while about his experience, but he had little to add. His testimony, however, rang so many bells with me and my own experiences earlier that year at Bolam Lake that I could not help feeling a warmth of kindred spirit towards him. Benny wandered off to join his girlfriend at the crap tables, and I rejoined my pals and the bevy of showgirls at the other end of the bar. I had a lot to think about, but later that evening yet another piece of the puzzle fell into place.

I, too, remembered that particular episode of *Fireball XL5*. Set in the years 2062/63 the series features the spaceship *Fireball XL5*, commanded by Colonel Steve Zodiac of the World Space Patrol. Also aboard as part of the crew were the glamorous Doctor Venus, middle-aged navigator and engineer Professor Matthew Matic, and Zodiac's co-pilot Robert the Robot, notable for being transparent. Robert was also unique as the only character in an Anderson series that was actually voiced by Gerry Anderson himself, albeit with the aid of an artificial larynx.

The series ran from 1962-3 and I believe that the episode that Benny and I remembered was called 'Faster than Light', and was first broadcast in October 1963. An Internet fan site describes the episode thus:

```
While en route for Space Station 9 to deliver much-needed supplies, Fireball
XL5 suddenly goes out of control. Despite risking death from radiation expo-
sure, Matt enters the ship's atomic core to attempt repairs; however, as the
ship continues to increase its speed the professor realises that there is no
hope. A chain reaction builds, and as the acceleration causes Steve, Venus and
```

Island of Paradise

Matt to black out, *XL5* breaks the light barrier. When the crew finally regains consciousness, they find themselves billions of light years away, emerging in a sea of air…

As I remember it, this is where the gallant crew encountered the mysterious giant birds, which both Benny and I remembered so vividly. However, my quest for the truth will go only so far! There are no screen shots of this episode to be found online, (or not that I could find anyway), and whereas the marvellous YouTube includes some XL5 episodes, it doesn't show this one. So, as I can't be bothered to actually spend thirty quid on something I don't actually *want*, and even the BitTorrent engines have failed me, then I shall do what any good journalist does, and make my excuses, and give up!

At every hotel-based UFO convention that I have ever attended, somebody's room becomes designated as party central. In my earlier days it was usually mine, but even at the height of my reputation as a party animal, I soon became tired of being unable to perform my ablutions because either there was a mountain of crushed ice and bottles of beer in my bath, or a total stranger throwing up in the toilet. Many years ago I decided that it was highly unwise to allow other delegates at UFO conventions to know my hotel room, and I have usually managed to survive these events with my sanity just about intact, purely because I have a bolt hole to which I can retreat away from the madness of conventioneering. Birds have their nests, and the beasts of the fields have their burrows, and Jonathan Downes - the Director for the Centre of Fortean Zoology - makes damn sure that he has a quiet refuge where there are always Earl Grey teabags, Jack in the Black, and Gram Parsons on the CD player.

Las Vegas is one of the most appallingly sybaritic places that it has been my misfortune to visit. It is such a disgustingly opulent waste of human resources that I shudder whenever I think of it. I dislike going there, and always feel that while I am visiting I am in the lair of what Osama Bin Laden calls - with some justification - the 'Great Satan'. I wish that US convention organisers didn't always insist on holding their bunfights in such a disgusting environ, but there is nothing I can do about it, and as a large part of my continued livelihood involves appearing at such events, I grit my teeth and bear it. When it Rome you do as the Romans do, and when in Vegas you party.

At this particular convention there were three party centrals. At the top of the social ladder there was a very W.A.S.P.ish event run by the conference organiser, Mr. Ryan Wood. Here could be found the elder statesman of American UFOlogy together with a gaggle of ageing capitalists with too much money and too little sense. I had visited this party once, but had blotted my copybook when one of the aforementioned ageing capitalists, after telling me how he had single-handedly won WW2 for Britain, tried to inveigle me into a political debate. *"We all really admired your Mrs. Thatcher",* he began (which was a statement which could have been designed to raise my hackles), *"and we admire your Mr. Blair even more."* My hackles rose even further, and I felt an irresistible urge to behave badly. *"Wouldn't you say that your Falklands War was the British equivalent of Vietnam?"*

"No dear boy, we won", I said with as much of a haughty patrician air as I could manage, invoking every generation of Englishmen who had sneered at the colonials ever since they ruined some perfectly good tea in Boston harbour.

I wasn't asked back.

The second party that I visited was so resolutely downmarket that even I was appalled. I stayed there just long enough to see the semi-naked go-go dancer with tassels on her nipples gyrating in the middle of a double bed, across a room whose atmosphere was so thick with marijuana smoke that conversation was nigh on impossible. To cap it all, they were listening to the *Spice Girls*. Like Kenn and Peter that previous afternoon in the hotel bar, I made my excuses and left in search of a party, which would hopefully be neither rarified and pompous, nor completely degenerate.

I found it unsurprisingly in Greg Bishop's room. Greg - an engaging young man of Japanese-American ancestry - was close friends with Kenn Thomas and wrote books that occupied much the same area of the Fortean spectrum. Together with my ex-wife I had once spent an evening with Kenn Thomas and his British counterpart Robin Ramsay, and I remember saying to her *sotto voce* that they were both lovely people, but after an hour at the same table with them I didn't give a toss *who* had killed Kennedy. However, the party in Greg's room was a great success. Nobody tried to patronise me, and there were no class A drugs in evidence, and a good time was had by all. Somehow through the bush telegraph which always seems to operate during UFO conventions, the word had got out amongst the most interesting conventioneers that *this* was the place to be, and during the course of the next several hours I met some interesting people, made some delightful new friends, drank a modicum of ice cold beer, and had a generally convivial time. Towards the end of the evening, an elderly man - with college professor written all over him - came up to me. He was holding a bundle of expensive looking large format hard backed books under his arm.

"I enjoyed your talk this afternoon Mr. Downes", he said to me *"and I thought you would like to see these"*.

They were a collection of books on Hispanic religious art. What this chap was doing at a UFO convention I have no idea. Furthermore, what he was doing carrying these remarkable tomes around with him, I have also no idea. But over the course of the next 40 minutes, he showed me that all my preconceptions about Latino demonology were completely wrong.

Like so many researchers, when I had been told that the chupacabra had "the face of the devil", I had conjured up a mental vision of depictions of the hornéd one that appear in most western European art from the Middle Ages onwards. They portray Satan as having the horns of a goat, a long pointed face with a goat-like beard, and reptilian eyes.

Like most students of contemporary neo-paganism, and images of the occult, I was aware that the medieval artists of western Europe had - on instruction by the Church - deliberately confused the Judao-Christian iconography of the Father of Lies with Cernunnos - the goat-honed pagan fertility god who - often paralleled with Pan - was the major male deity of the old religion of much of Europe before Christianity became prevalent. Despite knowing full well that these depictions were both cultural and daemonologically inaccurate I still fell into the trap - as have most of my peers - of assuming that when a Puerto Rican peasant described an animal he had seen as having "the face of the Devil", that it meant that the chupacabra had a face like that of a snub-nosed goat.

Not so.

My new acquaintance showed me page after page of paintings, engravings, and ritual masks from religious festivals depicting the Imps of Tartarus as depicted by Latin American artists. Without exception they were rat-faced entities, often with scaly skin and bat-like wings - but unquestionably rodentine in appearance. It appeared that during all my years in search of the chupacabra, my mental image had been of an entirely different beast. The famous drawing of the beast cobbled together I believe, by Jorgé Martin in about 1996 had been taken from eyewitness accounts and had given the creature the face of a western European devil. Maybe all of us have been looking for the wrong thing ever since.

By this time I was getting heartily sick of Las Vegas.

I wanted English tea, English sausages and English television. I wanted my own bed, back in my own little house in Exeter. I said my farewells and went back to my room where I made myself a cup of tea and lay on my bed. I pressed the play button on my CD player and totally by chance, the Georgia Peach came through the speakers singing his very own hymn of despair about the town in which I found myself:

> *This old town is filled with sin,*
> *It'll swallow you in*
> *If you've got some money to burn.*
> *Take it home right away,*
> *You've got three years to pay*
> *But Satan is waiting his turn*
>
> *This old earthquake's gonna leave me in the poor house.*
> *It seems like this whole town's insane*
> *On the thirty-first floor your gold plated door*
> *Won't keep out the Lord's burning rain*
>
> The Flying Burrito Brothers:
> *The Gilded Palace of Sin* (1969)

"Too bloody right", I said to myself, took a sleeping pill, and drifted into a fitful and uncertain sleep. But we will return to my ill starred sojourn in Las Vegas, and my motivation for doing it in the first place, later.

Six months later I was back in England doing what I usually do these days, which is attending to the incredibly tedious minutiae of running the world's biggest cryptozoological research organisation. It always mildly amuses me when I am at conferences or on the road because more often than not someone will come up to me and compliment me on having such a wonderful and exciting job. Of course it is wonderful and exciting - some of the time anyway, but as we get bigger and better at what we do, the administrative tasks increase exponentially and half the time these days, a visitor the CFZ offices would find it manned by two irritable and overstretched middle-aged men with beards doing their best working out how to balance the books at the end of the month, whilst still financing our programme of research publications.

July 2004 was a particularly stressful time anyway. After about 18 months of plain-sailing we were undergoing the beginnings of a major cash crisis.

Richard had just returned from his second expedition to Sumatra. It had cost quite a lot of money to mount, and we were dismayed - though not wholly surprised - to find that nobody within the national media was even the slightest bit interested in buying any of the pictures that the boys had brought back. From the scientific point of view the pictures were fascinating - they showed trackways through the jungle, vegetation broken as if by a tiny biped, and even the photographs of the valley into which no European explorer has every been before. However, at the time the media were engrossed in yet another sex scandal within the Royal Household - or was it the England football team? One editor of one of the less salubrious tabloids told me that if we had managed to photograph, or better still have brought it back alive, he might have been interested. I told him that if we had managed to make scientific history we would have been asking a damn sight more than the £300 he was prepared to pay for the story. He told me to complete a biological impossible act of self-impregnation, and slammed the phone down. *"You too mate"*, I said and lit another cigarette.

THE FACE OF THE DEVIL: This ritual mask from Oaxaca is a "Negrito Colmilludo" (tusked little black one). The Baile de los Negritos Colmilludos is unique to the "Villa Alta" sierra near the historic pueblo of Yalalag and the dance is well documented in paintings dating from the 19th century.

The party in Greg Bishop's room. Note complete absence of debauchery. Together with the author in the top picture is Greg (centre) and Peter Robbins. On the bed below is Jim Moseley. I have no idea who the others are.

The author and Nick R lurking in the desert

The author holding the Florida footprint, summer 1999

Island of Paradise

A TV company told us that they would be prepared to come along and film our next expedition for a documentary if we covered all their expenses. In return, we would have a 10% profit if, and only if, the show was sold to one of the major networks. I didn't even bother to dignify that suggestion with an answer. In an era where the mass media are obsessed with reality TV, spurious self-styled celebrities who are famous for only being famous, and sex lives of sporting heroes, cryptozoology has a hard time surviving.

To make matters even worse, the trusty CFZ computer had somehow contracted a series of unpleasant viruses and the Windows OS was at death's door. Never more than at this time, I managed a wry smile when thinking of the person who said Microsoft Works was an oxymoron. It was late on a Tuesday evening, and Graham was surrounded by various pieces of computer equipment as he desperately tried to salvage as much of our archives as possible. We had backups of course, but as always seems to be the case when the system crashes, the backups weren't properly up to date, and for three horrible days at the beginning of July, it looked as if the most recent versions of my book, *Monster Hunter*, Richard's book *Dragons: More than a Myth* and my father's two *magnum opii* had disappeared into the aether.

Then the telephone rang.

For some years we had been helping to finance the CFZ by doing contract work for various magazines and publishers, and it was not unusual to receive phone calls from my clients at socially undesirable times of the evening. I picked up the phone and when I heard that it was the unmistakable nasal voice of one of our biggest clients, I swallowed my frustration at the ongoing computer problems and did my best to be civil to her. I needn't have bothered; she was phoning up to sack us.

There was no need to go into details here. Indeed it would be grossly unprofessional for me to do so. Sufficient to say that this client had fallen out with one of our other clients, and had decided to throw the baby out with the bathwater by dispensing with our services as well.

I put the phone down, angry and upset. We had been extraordinarily loyal to the bloody woman when she had been in a cash flow crisis some months earlier. We had worked long hours, and waited for months before being paid. And this was how she repaid us! No sooner had I put the receiver down that it rang again. It was someone trying to sell me life insurance. Reacting unpleasantly to my repost that the only thing I needed insurance from was insurance salesmen, he slammed the phone down.

I was half way through telling Graham that, on top of the computer problems we were having, we now had to look to someone else to pay us £400 per month, when the phone rang again. I picked it up and answered in a less than relaxed fashion. It was Nick Redfern - now an ex-pat Brummie - phoning all the way from his new home in Texas.

"Aye up Jonny", he said, *"how do you fancy going to Puerto Rico for a week?"*

I gasped. What on earth could this be about?

I asked him when. He replied, *"Next Thursday"*.

I sat back into my chair in amazement. A very dear friend of mine, David Curtis, once told me that when he was a little boy his grandmother had always said that he shouldn't worry about what happened tomorrow because something would always turn up. In recent months I had started to invoke the old lady as one of the patron saints of the CFZ. She had come up trumps today!

Island of Paradise

It turned out that Nick had been working as a consultant for a company in Los Angeles called Cosgrove-Meurer Productions, who were contracted to make a series for the Sci Fi Channel in the United States. The series was called *Proof Positive* and its *raison d'etre* was to examine scientific proof for claims of paranormal or mysterious activity. Months before, Nick had asked me for some suggestions for the show, and off the cuff I had suggested that the TV company could get scientists to examine the feather samples that I had brought back from Puerto Rico in 1998.

Although the carcass of the chicken that I had examined back in the tawdry little farmyard in Dorado had been dry, desiccated and brittle, many of the other chupacabra reports - including some that I had investigated personally - reported that the puncture wounds on the neck of the victim had been covered in a weird glutinous slime. Some people have suggested that this slime is somehow akin to the anticoagulant, which is secreted into open wounds in the saliva of vampire bats to facilitate their macabre feeding methods. Although there had been no sign of any such thing at the farmyard, and indeed Oscar and George - the farmer and his mate - had no memory of any of the deceased roosters being afflicted in this manner, it did seem like a reasonably good idea to have the feather samples tested.

However, on my return to the UK I spent so much time collating the evidence that I brought back from Latin America and gathering new information which had been sent to me in the meantime that finding a laboratory to analyse these feathers for traces of the chemical which probably weren't there anyway were not exactly top of our list of priorities. However, if someone else was prepared to pay for it ....

It turns out that Cosgrove-Meurer were indeed very interested. They asked to see some of the footage that we had taken whilst in Puerto Rico for the first time. Praying that Graham would manage to fix the computer, I told Nick that we would get some stills and a sample of footage showing me examining the dead bird eMailed to the TV production office in Los Angeles within 24 hours.

"What the bloody hell did you tell him that for?" said Graham angrily. He had been working on trying to get our computer operational again for three twelve-hour days straight, and was nearly at the end of a not very short tether. I apologised but explained to him that this was probably the only way that we were ever going to get a CFZ presence back on the island of Puerto Rico in the foreseeable future, and if we didn't somehow pull a metaphorical cat out of the bag, then there was no likelihood of us being able to complete the research we had started so many years before.

"Well you had better make me a bloody cup of coffee then. Hadn't you?" Graham growled menacingly, *"and put some brandy in it. It is going to be a late night".*

Graham worked through until dawn the next morning. By this time the computer was just about functional. He snatched a few hours rest on my sofa, and continued work at lunchtime. By mid-afternoon we were able to make the first tentative frame grabs, and true to my word by 9 o'clock that night, we had eMailed the relevant *. mpg files to Cosgrove-Meurer in Los Angeles.

Graham went home late on the Wednesday night with a promise that he was going to sleep at least until Friday, and that we were unlikely to see him again until the beginning of the following week. I told him that I was incredibly grateful for everything that he had done, and said that I, too, intended to sleep for as long as I possibly could.

Ever since my mother was in her final illness in 2002, I have had a telephone by my bed. Usually, after a long day, I put the answerphone on and do my best to sleep the sleep of the just but on this occasion I forgot. I was woken up at the unholy hour of five to seven. It was a researcher from Cosgrove-Meurer. He had miscalculated the time difference, and in a misguided act of kindness stayed up incredibly late in order to phone me at what he thought was a sensible time. He was so enthusiastic and ingratiating that he reminded me of my old dog Toby back in the years when he was developing from puppyhood to the state of full-grown dogginess. In the middle of the night he would go down to the kitchen, turn out the rubbish bags, and stealthily bring up a pile of heavily-masticated and misshapen baked bean cans which he would leave at the bottom of my bed as a present for me when I woke up. I am not a good sleeper, and not at my best first thing in the morning, and when the first thing that you see when you open your eyes is the aforementioned array of household waste, it is only the fact that there is a little black and white dog with wide doggy eyes looking so pleased with himself at the treat that he has prepared for his master that saved him from a fate worse than death.

Despite the fact that it was an unholy hour of the morning - I hadn't even opened my eyes, let alone had a cup of tea before having to launch into an impassioned discussion about vampiric attacks on livestock on a small island in the Caribbean. As the researcher spoke, I imbued him with big doggy eyes, fluffy ears and a waggy tail, and forgave him for his unpardonable act of social blasphemy.

Yes, they were very interested in having a forensic team examine the feathers, but they wanted more. They wanted me and Nick to go to the islands, interview new witnesses, and talk about the history of the chupacabra phenomenon. Was I prepared to embark on the Thursday of the following week? Of course I was, but having been badly bitten in the past I wanted to make damn certain that I had established my position within the project once and for all.

I explained to my new friend that whilst I was a scientist who did his best to work within the strictures of scientific methodology, I was somewhat unconventional both in my outlook and my appearance. Although I have lost quite a lot of weight in the last few years, I still have what the big and tall man's shop in South Devon describes euphemistically as a fuller figure, and, as a result of various physical mishaps and ailments with which I shall not bore you, I walk with a stick. I was not prepared, I told the researcher, to take part in any filmic sequences that made fun of my weight problems and disability. He was appalled at the very idea. *"Of course not, Jon. We wouldn't dream of it......"*

I also wanted to make sure that I would be given enough film time to explain both my theories and the methodology behind them. I was assured that there would be no problem.

Having reached the age where, having flowing locks no longer means I am an antisocial member of society, I usually wear my hair longer than my academic contemporaries. Would this be a problem? Nope, I was assured. They knew perfectly well what I looked like, they had read my books, they had seen me on other television programmes, they knew what they were getting and they liked what they saw. Would I be prepared to sign a contract to make a show for them? How could I possibly refuse? He agreed to eMail me the contract, and after a phone conversation that had lasted about 20 minutes, I was free to go back to sleep.

Of course, I was too excited to do any such thing.

I went downstairs, made myself a cup of tea and took 'Tessie', the CFZ dog since 2000, out for a leisurely stroll. When I returned, the rest of my household was still fast asleep, so I sat down at my trusty computer - now restored to full working capability - and began work on my preparations for the forthcoming trip.

It is ironic that Nick Redfern first made his name in the public eye with a book called *A Covert Agenda* because this was precisely what I had. Although I was quite happy with the concept of continuing researches into the chupacabra, as far as I was concerned, the mystery was probably unsolvable. The chupacabra was just too weird to fit into any accepted zoological frameworks, and years before I had regretfully consigned it to the realms of paranormal activity that was probably never going to be solved in my lifetime.

However, I had other fish to fry on the Island of Paradise, and whilst there was no way I was ever going to be able to convince a TV company to fund my researches into what I really wanted to work on, I was pretty sure that if they flew me out to make another chupacabra documentary, I could find the time to do a little sleuthing of my own.

Richard is a late riser. He got up at around the time that most people have lunch, and I told him my good news. Together we started packing my bags with tropical clothes and collecting equipment that I would need. It all seemed too good to be true, and in my experience when something seems too good to be true it usually is. There was bound to be a fly in the ointment, but at that stage I was damned if I could see what it was.

Halfway through the afternoon the phone rang again, and it was someone more senior in the management of the film company. He dropped a bombshell. "They" (and I have to admit I am not quite sure who they were) had decided that, as well as the feather samples, they would also need a plaster cast of the chupacabra footprint which I had - according to them at least - in my possession. The problem was, I didn't have a chupacabra footprint. What I did have was a plastercast of a footprint that had been given to me in Miami during February 1998 by a well-meaning UFO expert who claimed that it had come from a chupacabra. The problem was that neither the footprint, not the animal that the eye-witnesses who had originally procured the footprint claimed had made it, had anything to do with the chupacabra.

The footprint had been taken from a pile of fine builders' sand outside an old peoples' home in Sweetwater, Florida in 1996. The matron in charge of the home had heard a commotion, and upon looking out of her office window had seen a strange creature that she described as being half-way between an ape and a dog chasing a chicken across the yard. It had killed several of her fowls, and by the time she had gone outside to investigate, the creature had disappeared, leaving carnage in its wake. The plastercast, which was taken later that day, was a singular object indeed. It appeared to be of a large carnivore but the strangest thing about it was that, instead of claws like a dog or a cat, it had fingernails like an ape or a man. Ironically, the advent of this strange creature had taken place at the height of a spate of UFO activity in the region whilst - on the other side of the Atlantic - at the height of another spate of UFO activity, a strikingly similar footprint had been filmed in Rendlesham Forest, Suffolk, by my friend Maxine Pearson. Interestingly enough both regions had folklore relating to strange ape-like dogs (or possibly dog-like apes).

In Florida and some of the other south-eastern states of the USA there are reports of a chimpanzee sized creature called the skunk ape. Less well known than bigfoot, the accounts are even more believable. Veteran Fortean researcher Loren Coleman has suggested that they are a descendant of a Miocene ape called Dryopithecus or oak ape. It is a nice theory, but sadly there is a dearth of fossil evidence to support it. There are no non-human higher primate fossils in North America, and if there is a zoological explanation either for bigfoot, the skunk ape or any of their kin, it must have slipped through the gaps in the fossil records.

The skunk ape is often described as being like a cross between a large dog and an ape, and is reported to smell foul - hence its name. Less odiferous is the shug-monkey of medieval Suffolk folklore. This creature is a demi-daemon of the woodlands of East Anglia and has been reported - often in conjunction with a strange light in the sky - intermittently ever since.

Island of Paradise

For many years I assumed that what I had in my possession was indeed the footprint of a skunk ape. I touted it around conventions, and was always pleased with the enthusiastic response that I got from people when I pointed out the dog-like pugmarks, and the ape-like fingernails. However, in early 2003 I was sent a series of photographs of plastercasts of a known creature - a jaguar - that had also been taken from fine builders' sand, and I was shocked (and somewhat embarrassed) to see that as a result of the action of wet Plaster of Paris being poured into a semi-porous surface, claw marks could soon become misshapen and appear like fingernails.

Although I am loath to discount the matron's evidence of what she saw, my footprint would not stand up in a court of law. Although they may be of an unknown species, they are more likely to be of a known albeit very rare one. The Florida panther is a highly endangered sub-species of the cougar, puma or mountain lion. There are only a very few left but the outskirts of Sweetwater are well within their range. I think it more than probable that what the matron saw was a female of one of these extraordinary rare big cats 'teaching its young to hunt'

Whatever the footprint is, it is certainly not that of a chupacabra. People have reported seeing chupacabra footprints but they have much thinner pugmarks, with elongated toes or claws - a bit like those of a hamster writ large. I was not minded to appear foolish on television by presenting a piece of evidence, which had absolutely nothing to do with the case that we were supposedly trying to solve. As I explained to the executive, it was of a different animal from a different country, at a different time. Whatever it was, it had nothing to do with the case. Any road, the plastercast was a definite one-off and I was highly unwilling to entrust it to the tender mercies of a TV production company - or an airline intercontinental service for that matter. Not only would it add nothing to the programme, but also it would put one of my more valuable props at risk. I was not going to let them use it.

He was disappointed with that. However, he appeared to take it in good grace and then dropped the second bombshell. They wanted the feather sample and the original videotapes of the 1998 expedition as soon as possible. I had no problem with sending feather samples off, but was unwilling to let my master tapes go without having agreed a heavy indemnity from the company if they were to be lost or damaged. We haggled for about ten minutes and we eventually reached an agreement that my master tapes would be insured for the hefty sum of $10,000 apiece. I telephoned Graham to tell him. *"Let's hope they destroy the bloody things"*, he grunted. *"We need the money!"*

Ten minutes later the phone rang again. It was Nick Redfern sounding horribly embarrassed. *"Er, Jonny there is a problem"*, he mumbled. For some reason, when he is under stress, his natural West-Midlands intonation, which has been tempered somewhat over the years since he has lived in Texas with a southern drawl, come to the fore and he sounds Brummier than ever. *"Er, the wazzocks say they say they won't make the film if you don't send 'em the footprint"*.

I will draw a discreet veil over what happened next.

I knew that it wasn't Nick's fault, and that he was merely in the appalling position of trying to persuade one of his best friends to do something that he would dislike intensely, at the behest of the TV producer who he was beginning to despise. I don't like being bullied, especially not by an idiotic TV producer who was trying to use one of my best friends against me. However, both Nick and I knew that I really wanted to go to Puerto Rico again, there was very little we could do, and that eventually I was going to have to give in. However, I told Nick that he could tell the bloke in Los Angeles that he should not ask somebody else to do his dirty work for him, and that he should phone me directly. Nick giggled, and told me that he had already suggested that, but that the producer was scared of me.

"So he bloody should be", I grunted. Nick laughed, and phoned the Los Angeles office once again.

Before I had had a chance to finish my cup of tea a chagrined TV producer was on the phone again. With the worst possible grace I haggled, and while Richard was sat in the other corner of my sitting room, a hankie stuffed in his mouth trying to stifle his rising hysterical laughter, I negotiated an insurance deal of $1,000,000 for a footprint which had nothing to do with the television show, and which had cost me precisely nothing. I then phoned Graham again.

"For God's sake can't you make sure it disintegrates in transit. $1,000,000 ....can you imagine how much lager that would buy?"

Then the phone rang again. It was my nemesis from Los Angeles. *"We want you to Fed-ex the footprint, the feathers and the video tape to us today"*, he barked. *"Mmmm there is one problem with that, dear boy"*, I said in the tones which I reserve for those moments when I want an irritating colonial to know that he *is* an irritating colonial. *"I live 250 miles from London and I have no idea where the nearest Fed-ex office is."*

"But it is the Federal Express!" he cried in a voice of wounded dignity. *"Every small town has them"*...

"In America perhaps", I said with a sigh *"But this is England. I can have them shipped to you tomorrow by carrier but it will probably take me several days to contact anyone from the Federal Express company"*..... He put the phone down saying that this was a matter beyond his remit, and he had to consult with his superiors. Soon after, Nick was back on the phone again.

"He says you won't bung them feather samples in the post", he said in a complaining voice.

"Not at all, Nicky", I sighed. *"I just couldn't make the bloody man understand that there isn't a Fed-ex office in Exeter and that this is the United Kingdom, not somewhere in the arse-end of rural Idaho".*

We both laughed. Now I would not want anybody reading this book to think that I am anti-American in any way. Although I am quite prepared to admit that what I have written in this chapter so far is probably enough to have made President Bush to put me on his list of the `Axis of Evil`, I actually admire America, I have a lot of fun there, and have many American friends. I just become frustrated with the parochial attitudes of some of the less enlightened of their citizens. They seem to think that England is a tiny medieval country where everybody is either a Cockney or a member of the aristocracy and I find it somewhat galling when I meet a perfectly affable American for him to say *"Oh gee, you're from England. Do you know my friend Tracy from Middlesborough?"* I try to explain without being patronising that there are 65 million of us and that it would take well over 24 hours to drive from Land's End to John O'Groats, but I can always see his eyes glaze over as he imagines that the United Kingdom is pretty much as was depicted in *Mary Poppins*.

I said as much to Nick, and we both grinned. He rang off and phoned Los Angeles again, only to phone back within minutes saying that the company apparently insisted on using Fed-ex. Eventually I found a Fed-ex office in London and arranged for them to come and collect the parcel at the beginning of the following week. The real irony was that this meant that I would actually be in America only about 48 hours after my parcel, and if they had been prepared to wait an extra few days things would have been much easier. However, the saga was far from over.

Fed-ex refused to take the parcel.

Apparently feathers were now classed as biohazardous material under strict US Government legislation brought in the wake of the anthrax attacks, which followed soon after the events of September 2001. Whilst being irritated by this, I could see their point, and especially considering that the whole point of the exercise was to analyse these feathers to see if they did indeed carry some obnoxious chemical, one could quite understand the carriers refusing to take them.

So I telephoned Redfern again. I told him to assure the little man in Los Angeles that I was not being deliberately obstructive, but told him how, not only had I been told that they refused to take feathers but that the man in the London office of Fed-ex had said that if I sent the feathers incognito as it were, and was found out, I would be liable for prosecution under American law for committing bio-terrorism. Even during my anarchic youth I would never have gone *that* far to cock-a-snook at society, and - as I told Nick - the prospect of being the first cryptozoologist in Guantanamo Bay did not appeal.

I then had yet another call from Los Angeles. They asked me just to slip the feathers between the pages of a book and send them. This I flatly refused to do. First it would ruin the whole point of the exercise. If there was indeed a thin coating of alien chemicals on the feathers, they would have no doubt had been disturbed by the process. Six years before I had taken care to make sure that the feathers were kept under as hermetically sealed conditions as possible, and I wasn't going to spoil it now.

Secondly, quite apart from the legal implications (and by this time I had discovered that I would be automatically extradited to Washington to stand trial if I was found out), there was the outside chance that there was *indeed* a noxious chemical on the feathers, and there was a very real - if slim - chance that by committing such a foolhardy act I would actually *be* guilty of bio-terrorism. After about ten minutes pointless argument, during which I agreed to send the footprint and video tapes by Fed-ex, and to take the feathers with me in my hand luggage still in hermetically sealed test tubes but in no other circumstances, they finally agreed…..with the worst possible grace.

The human tragedy of September 11th 2001 had reverberated across the world. It was - like the deaths of Kennedy and Lennon - one of those world-shattering events which one knows will be remembered for a lifetime. There will be few men or women on the planet who will not recollect for the rest of their lives where they were when it happened. I was recovering from 'flu and lying in bed that afternoon. My friend Tim Matthews - a notorious UFOlogist and author who has featured elsewhere in my writings telephoned me. *"Switch the TV on now!"* He barked. I did, and watched in awe as the television showed endless repeats of the planes crashing into the towers of the World Trade Centre.

It had changed the world forever in many ways, but one of the ways that has largely passed unnoticed from the public eye, was the way it has affected organisations like the CFZ.

Before 9/11 it was reasonably easy to go to Gatwick or Heathrow and wait for a cheap stand-by ticket to anywhere you wanted to go in the USA. This facilitated cheap and easy foreign explorations. However, in the years following the tragedy of the World Trade Centre, legislation has tightened up, and the whole process became considerably difficult and more expensive. Security had been tightened up on all levels, and the process of foreign travel - especially to the United States - was not anywhere as near as much fun as it had been a few years before. Only a few months previously, when I had flown to Las Vegas for the UFO Conference described earlier in this chapter, I had been searched, had to turn out my luggage, and even had my walking stick taken away at Gatwick Airport by officials from American Airlines who seemed convinced that I was a threat to national security. OK, the day I flew out was the same day that George W. Bush had flown into London for a state visit, but it didn't make me look forward to the forthcoming journey. But, the die was cast. I was going back to Puerto Rico and despite any irritations and difficulties I was going to be able to carry on with the research that I had started several years before.

Englishmen abroad: The author and Nick Redfern, Las Vegas, December 2003

The Margarita Diary

Chapter 2 Comments:

Having digested Chapter 2 of Jon's mighty tome, I can safely say that one thing stands out more than any other: only an adventure involving the Centre for Fortean Zoology could result in a deep discussion of *Fireball XL5*, Earl Grey Tea, Guantanamo Bay, chupacabra DNA, and the ominous Department of Homeland Security!

But that's fine with me. If I wanted a humdrum life packed with boring, predictable souls, I would be working deep within the heart of the dreaded world of 9 to 5. And as the mighty *Ramones* so aptly put it: 'It's not my place in the 9 to 5 world.' And neither is it Jon's, I'm immensely pleased to say.

I'm well aware that the image that certain people have of the CFZ is that we either (a) make immense amounts of money, live like landed gentry and drive around in Ferraris; or (b) spend all our times out of our brains on booze and chemicals. The truth is somewhat different. Yes, altered states have not been unknown at times within CFZ circles; but Ferraris and ever-flowing pounds and dollars are not a regular sight.

Rather, we are a group of people for who every day is an adventure - but also for who, every day is an on-going battle to keep the wolf from the door, to thrive, to survive and to ultimately triumph - albeit in an unconventional fashion that suits our unconventional lives, and most important of all: completely utterly according to our own rules.

So, when I phoned Jon and asked him if he could fly to Puerto Rico 'next Thursday', it really was just another day in the unconventional world of the CFZ - and one that saw me busily handling the American side of things, Jon scrambling to make things happen from the UK, and Graham dutifully toiling day and night to ensure that everything that the Sci-Fi Channel needed, they got; where humanly possible, of course - and the automatons of the Department of Homeland Security notwithstanding.

In other words, we may be a slightly anarchic bunch at the CFZ, and we may be decidedly unconventional in many ways, but I am proud to say we know how to get the job done and do what is required – just as Jon notes in the above chapter describing the crazy build-up to that even crazier week on Puerto Rico. **NR**

Chapter Three
Snail hunting for fun and profit

As mentioned above, Nick Redfern is not the only bloke in this book who has a covert agenda. My main aim in revisiting the island of Puerto Rico was not to investigate the chupacabra. It was actually to go and search for something far smaller.

Many years ago, when just a callow youth, I went threw a phase of reading the books of people of the ilk of Eric von Daniken. The "God was an astronaut" brigade had a firm hold of the culture of the mid-70s, and as a spotty schoolboy in rural North Devon I found his books mildly interesting. However, as the original tome *Chariots of the Gods* gave way to innumerable sequels, I began to smell an enormous rat! The original thesis; that mankind had been visited in pre-history by intelligent beings from another galaxy who tinkered with our DNA and created us "in their own image" was mildly interesting though didn't really hold water. As the decades have progressed, the author tried to link in so many other earth mysteries and historical anomalies that the whole premise became mildly ridiculous and no-one with any real degree of self-awareness took him seriously anymore. He finally blotted his copybook for good at the end of the decade when at least one statement he had made about his activities in South America proved to be palpably made up, and tainted the rest of his evidence irrevocably.

By this time there were so many books - by him and many other authors - that one began to expect that the next episode would be Sid James *et al* appearing in *Carry On Chariots of the Gods*. One could almost imagine him holding up a model flying saucer next to Barbara Windsor's cleavage, and cackling something about *"you don't get many of those to the pound do you missus?"* In one of these books, however, I read something that stuck in my mind for many years. It was about a visit to some caves high up in the mountains of Puerto Rico. On one of these cave walls there was, so he asserted, a carving of a strange spinning-top shaped object which, he insisted, had been done by the Taino Indians back in the mists of time, and which he claimed could show nothing else but a flying saucer. Here was irrevocable proof that mankind had, indeed, been visited by little green men from the planet Zog.

Like so much else, this piece of information had been sifted into the back of my mind, and although I had never completely forgotten it, to this day I cannot remember which of the books originally contained it. However, in 1998 when visiting the cave systems high above Aguas Buenas in the middle of Puerto

Island of Paradise

Rico, I was suddenly shocked when I found - if not exactly the same cave carvings, startlingly similar ones. For a few brief moments I was transported back to being a wide-eyed schoolboy in 1972 - one who actually believed in the ETH, and was bored enough with his maths lessons to read any old crap in a desperate attempt to try and escape.

Luckily, for my sanity (and for my reputation as a UFOlogical sceptic), within five minutes of discovering the carving, I had also solved the mystery.

There, crawling slowly up the slippery limestone walls of the cave, sliding up what would be in a century or two, an impressive range of stalactites, was a large and singular snail. I say singular because it looked just like one of Eric von Daniken's flying saucers.

What happened next is documented in my book, *Only Fools and Goatsuckers*. I discovered that the snails (there were several of them) appeared to live a most un-snail like existence in almost total darkness eating guano (bat poo to the uninitiated), and although I had been unlucky in my attempt to bring them back alive to England, I wanted dearly to try and do so this time.

Firstly, because in the intervening six years I had never been able to find out what species they were. Although a long shot, I would have not been surprised to find out that they were something entirely new to science, and I was interested to try and find out what they were. If they *were* a new species then their discovering would be a minor coup, both for me personally, and for the CFZ.

My second reason for wishing to bring a viable breeding population of Puerto Rican cave snails back to England was to try and prove my point over von Daniken's ludicrous notions of flying saucers visiting Puerto Rico. Although, as this book will show, I have in fact taken quite an interest in UFO activity on the island, and there has been a long and interesting series of such accounts from the days of Columbus on, I looked forward to being able to disprove at least one piece of UFOlogical hokum.

My third reason was purely mercenary. They were weird looking creatures, and I hoped if I could breed them I would be able to sell a few at vastly inflated prices to collectors of invertebrates.

For several years I have been forecasting that the next great trend in British pet-keeping is going to be the keeping of tropical invertebrates. As house prices rise, and with the increasing urbanisation of our sceptred isle, more and more people find themselves unable to climb on to the first rungs of the property ladder, or - if they do - they find themselves in small urban flats. In such places the keeping of dogs and cats is difficult, and is quite often forbidden by the terms of the lease. The past ten years has seen a significant rise in the number of people who keep tropical fish, and until just before the end of the millennium an increasing number of people were keeping tropical reptiles as well. However, the animal rights brigade had successfully put the kibosh on a lot of the reptile keeping community. There were no more reptile fairs in Britain, despite the fact that the vast majority of people who keep exotic pets are real enthusiasts and even scholars, rather than just ordinary people who want something weird in a tank. The animal rights lobby had successfully evoked archaic laws designed originally to stop farmers selling livestock in unlicensed premises, and using the pretext that such events were cruel to animals (which they weren't), in one fell swoop the major forum for enthusiasts to buy, sell and swap exotic reptiles and, possibly more importantly, the major opportunity for enthusiasts to meet each other, was closed forever. They then started to attack the reptile shops. Whereas once upon a time there were dozens of specialised reptile shops; any major city would have a couple, now there were very few. Incessant picketing and harassment had forced one reptile shop owner - a female acquaintance of mine in Middlesex - to commit suicide. Despite the fact that reptile hobbyists were carrying out a great deal of valuable scientific research, and in many cases taking part in important multi-national breeding programmes, the hobby had received a body blow from which it would probably never recover.

I was convinced - and still am - that exotic invertebrates (who had not under any circumstances been covered by the legislation which governs the trade in more complex animals) would be the next growth area in pet keeping. Indeed, whereas twenty years ago one could occasionally buy stick insects from pet shops, now several dozen species of phasmid were commonly available, and animals such as African land snails were commonly kept as pets. Always looking for new and interesting ways to finance the CFZ, I felt that if I could introduce a hitherto unknown, but undeniably attractive (in a weird kind of way) animal into the UK marketplace, then I would most likely be able to swell the coffers of the CFZ in quite an entertaining manner.

So, how was I going to bring them back?

Under British law there are very few pieces of legislation that cover the import of carnivorous invertebrates. The law is strict when it comes to the importation of any animal that could become a pest to crops or livestock, but an animal which ate bat shit was, I thought, fair game.

However, one is not allowed to take snails into the United States. The law is strict on that count, and one has to fill in a Customs waver form as you enter U.S airspace stating that you were not doing anything of the kind. However, I was not going to be taking them *into* the United States, merely taking them out again, and I hoped against hope that there would be no such Customs checks when entering the mainland from Puerto Rico.

I was fairly convinced that on my previous attempt the only reason I had been unable to bring the snails back alive to the UK, was that we had visited Puerto Rico early on a trip which was to include visits to two other countries. I had been unable to find suitable food plants in Mexico, and I was sure that although the snails were still alive when I took them back through to Florida that the plants, which I had snaffled from the hotel garden in Miami, had been so heavily treated with pesticides that my poor creatures had been poisoned. Also, on the first trip, there had been the slight problem that I had no collection equipment, and had been quite unprepared for having to take care of two tiny dependents.

This time was going to be different. On the Monday morning before my flight (on Thursday) Graham and I went to the local supermarket and bought a number of large Tupperware containers. At the checkout the shop assistant smiled at me in a vacuous yet friendly manner. She could see from my other purchases that I was contemplating a trip abroad, and smiling said something silly about making sure I had enough picnic things with me. I looked blankly at her for a few seconds before realising that she had assumed that the aforementioned Tupperware was to hold sandwiches rather than a far more precious and arcane cargo. I wondered, for a moment, whether I should enlighten her, and then thought better of it. I grinned politely, paid the bill, and left.

On the Tuesday before I left, Graham and I decided that I was in need of a makeover. Whereas, during our first visit to Puerto Rico the CFZ website had been a tiny and primitive affair, by this time it was large, complex and most importantly, a valuable means of raising awareness about our activities and cash with which to carry them out. Having received an assurance from the researcher at the TV Company that I would be depicted in a positive and respectful manner, we sat down with Richard, and John Fuller - then, the most recent addition to the CFZ team - and had a brainstorming session. Whereas the Sci Fi Channel in the UK is a fairly minor-league affair, it's US sister station is markedly more important. They not only commission documentary series, but even the occasional feature film and I was painfully aware that this was an important chance for me - and the CFZ - to increase our public profile immeasurably.

For some years I had realised that my "Englishness" is a valuable selling point for me, especially when dealing with a transatlantic audience. I could not hide the fact that I am markedly overweight and in less than perfect health but felt that, with the appropriate skin, we could make these points work for me rather than against me. In recent years – in the UK at least - I have alternately sported a blue pinned-striped suit and tie, and a leather jacket and jeans, usually on TV wearing my hair in a pony tail when I wear the leather jacket, and letting my flowing locks blow in the wind when I wear my suit, giving the impression, as Richard once put it, either of a Columbian cocaine dealer or a yuppie version of Grizzly Adams. In the field I usually dress for comfort, and on our first sojourn in Central America I had worn jeans, lumberjack shirt, and a big voluminous denim jacket with poacher's pockets which was comfortable and convenient for holding dictaphone, camera and notebook (not to mention hipflask). However, I had not been happy with the way I appeared when immortalised on celluloid, and so was determined to take more care over my public image this time.

It was Richard and John Fuller who, between them, came up with image which I decided to sport for the Sci Fi Channel and which, ironically, I liked so much that I have worn in public ever since; a weird cross between Jimmy Page circa 1972, and the bloke in Graham Greene's *Our Man in Havana*. Graham and I drove down the little country lanes to the village of Beer (just outside Seaton in South Devon) where there is an emporium that I vulgarly refer to as the "Fat Bastards' Shop".

I have always disliked shopping for clothes, especially in High Street retailers. As I get older the lack of service, the appalling piped music which is supposed to be conducive to making one's customers parting with as much cash as possible, but which I find has the opposite affect, and most importantly that I can never find anything to fit me, has made my visits to such shops something akin to purgatory. I have to say though, that I really enjoy visiting Osborne's in Beer. Not only do they have a range of clothes which I find attractive, and which actually fit me, but they are reasonably priced, and the people who work there have a pleasantly old-fashioned and mildly deferential attitude towards their customers.

After an hour, Graham and I found ourselves driving back towards Exeter. I was nearly £200 poorer, but I had bought a white yachting jacket, three identical dark shirts, a panama hat of the type usually worn by cricket umpires, and had also been persuaded to spend a few quid on a new dressing gown, and some sandals. Later that afternoon, beard trimmed and resplendent in my new finery, Graham took me out to the Exwick Vortex for a photo session. It must have been a success, because seven months later a certain young lady in Lincolnshire went 'wow' when she saw the results.

However, the point of these promotional photographs was not just to make ladies in Lincolnshire go 'wow' but was really so, as the public face of the CFZ, I could make friends and influence people in strange new lands.

I think here I should add a few words about the Exwick Vortex. It is a piece of wasteland that is attached to the housing estate on which I lived from 1985-2005. When I first bought the house, we were told that the area was going to be a children's playground. In the intervening two decades it has become overgrown with brambles and assorted shrubbery and the only people who ever visited it were us. Over the years I have counted over a dozen species of butterfly living there, and it is one of the few places in the south of England where an exceedingly rare species of moth - the Jersey tiger - both lives and breeds. I have never actually seen any children there, and its main use was for us to allow the CFZ dog to answer her calls of nature and as a suitably neutral backdrop for publicity shots. Over the years we have been photographed there many times. It has doubled as Dartmoor, Exmoor, Bodmin Moor, the Peak District and even Roswell, New Mexico. Now - a year after we had taken the publicity photos for the first Sumatra expedition there - it was now the El Yunque rainforest.

This page and over: Our man in Havana …. er Exwick (July 2004)

Island of Paradise

The patch of wasteland has been known as the 'Vortex' for so long now, that I'm probably the only person left who remembers why. Back in the spring of 1997 Graham and I had been asked to contribute to a CD Rom which was to be cover mounted a on a well-known computer magazine. For the life of me, eight years later, I have only the slightest and vaguest memory of the subject of this compact disc, but doubtless it had something to do with the world of unsolved mysteries. Graham and I went into our usual schtick talking about Cornish sea serpents, and shamelessly plugging my book about the Cornish owlman. When we had finished, and the film crew had given us our 50 quid, we all sat down for a cup of tea, and they regaled us about some of the other things that they had had to film whilst making the CD. Apparently, only a few weeks before visiting us, they had been in Oregon at a place called the Oregon Vortex. From what I remember - and to be quite honest I was more interested at the time about how I was going to spend my 50 quid, so I wasn't really paying attention - this was a valley where compasses went haywire, and the normal laws of electrodynamics did not seem to apply. Graham - who is possibly one of the most peculiar people I have ever met, in a lifetime of meeting peculiar people - then claimed with a completely straight face, that the same phenomenon held sway in the tiny piece of waste land at the end of my path. And ever since, it has been known - by us, at least - as the Exwick Vortex.

I was very pleased with the results of the photo session, and found out quite how good the photos were the next day. We had posted one of them on the front page of our website together with a brief announcement that Nick Redfern and I were going off to Puerto Rico in search of the chupacabra. Within half an hour of doing so I heard from a jolly nice bloke at *Bizarre* magazine. Impressed both by my publicity shots and the press release which we had issued telling the world about our plans (and conveniently failing to mention that my real quest was for some two inch long snails rather than a four foot high vampiric alien) he wondered whether I would care to do an interview.

I don't mind doing interviews, but over the years I have done so many that I find myself saying the same things over and over again. This time was different. *Bizarre* being a "lads'" mag from the same stable as *Fortean Times*, but owned by the bloke who became a millionaire after FHM magazine a decade or so ago, they came at it from a completely different angle. He wanted to know whether my successes as a cryptozoologist had improved my successes with women. I laughed, and told him that sadly it had not, but this broke the ice and we had a very laddish and highly enjoyable chat. Upon my return, I was amused to see that they had printed my answers verbatim - expletives and all - and that the pillars of American academia that already considered me to be somewhat akin to an iconoclast, were unlikely to be impressed. However, the interview led to a great many other media appearances, and even got me another transatlantic journey in search of the chupacabra …

….but that is another story.

I always hate the time just before I embark on a journey. It is horribly reminiscent of those days in my mid-teens after a sojourn at home when I was preparing to go back to boarding school. My public school days were not the happiest of my life, but I hated the transition period far more than I hated living in the big communal dormitory and paying lip service to the *mens sana in corpore whatsit* value system that has irritated me ever since. It was the same as an adult. I quite liked my little house in Exeter, but then again I quite liked being away from it. It is the periods of transition when I am preparing to go, or when I am preparing to come back that irk me.

This was to be third transatlantic flight in just over six months, and for the first time I was doing it in style. Usually, in a desperate attempt to keep the costs down, I sleep all day at the CFZ and catch the one in the morning train to Gatwick or Heathrow. This is incredibly cheap, costing less than half what a normal train journey would, and allowing me to save the cost of a hotel room for the night, but it is a nasty journey, and can get quite hairy as it is the train on which the social undesirables seem to congregate.

Folk singer Roy Harper (a musician of whom I am quite fond) once wrote a song in which he bemoaned that whenever he took a journey he always found himself sitting next to the loony on the bus. On the late night train to London Airport, it was well nigh impossible not to do likewise.

So I insisted that those jolly nice fellows in Los Angeles pay for me to catch an ordinary train, and stay overnight in a hotel, before flying to New York for the connecting flight to San Juan. Much to my amazement, considering how difficult they had been when dealing with the matter of the feathers, the footprint, and the courier, they didn't bat an eyelid at this, and ever since I have been regretting the fact that I didn't insist on a first class ticket or even a limo to drive me to Heathrow from Exeter. However, there was nothing I could do about it, and so, at about half-past two on the Wednesday afternoon, Graham loaded me, my suitcase containing my clothes, my video equipment and my collecting things, a holdall, and my laptop into my trusty, though ever so slightly decrepit old Jaguar, and drove me to St. David's station. He promised to mind the office in my absence, plonked me on the train and (I suspect) drove back to the CFZ by way of the little off-licence.

I always find train journeys, and to a lesser extent, air travel put me into a state of limbo somewhat akin to that which I imagine that Schrödinger's eponymous cat would have felt in the famous experiment if it had ever taken place. I am in a state of complete indeterminacy, and whilst in the bad old days I would have probably whiled away a few hours having drinks in the bar, these days I don't do that sort of thing, and so I bought some provisions in the buffet car, and then ambled back to my seat, where I reflectively swigged at a bottle of diet Coke, ate a cheese sandwich, and opened my holdall to retrieve the book I had chosen for the journey.

I have written elsewhere about how my mother warped my literary tastes at an early age. Well before puberty I had been introduced to the late 19th and early 20th Century pulp fiction by people of the ilk of Edgar Wallace and Henry Rider-Haggard. I had recently - courtesy of the jolly nice fellows at eBay - purchased a copy of *Bulldog Drummond: His Four Rounds with Carl Peterson* written in the years following the Treaty of Versailles by a hack writer known only as "Sapper". I had adored the books when I was a child, and looked forward to reading them again. I soon found that, whilst they had their moments, and indeed even had somewhat of an S&M kitsch appeal, on the whole they were pretty hateful pieces of writing. Whereas the hero had enough good points to make him attractive, most of his followers were merely thugs, and the constant parade of villains (to a man, Jewish or German), were somewhat irksome. However, they were ripping yarns, and there was enough substance there to keep me happy for the next day or so.

Surprisingly the journey went quite quickly. Unlike other train journeys preceding foreign expeditions during which every second lasts ten minutes and every minute a decade, lulled away on the execrable prose of a long-dead racist, I found myself at Reading Station before I knew it. It is weird. I seem to have spent a large amount of my adult life on Reading Station and I know it like the back of my hand. When I was 17 I lived in Bracknell for about six months, and Reading was the nearest source of the fleshpots. In my adult life I keep on returning to the place, because either it is the jumping off point for the train to Gatwick or the coach to Heathrow, or it is where you change trains for Oxford, Salisbury, and a surprisingly large range of other places. I have visited it at all hours of the day or night and on more than one occasion, have had to sleep on the extraordinary uncomfortable seats in the waiting room on platform four. I had 45 minutes to wait for the shuttle coach to Heathrow, so I indulged in one of my favourite pastimes. I negotiated my way to the bus stop at the front of the sprawling edifice, sat myself down and watched the soap opera, which transpires whenever you observe the human race at any major transport terminal.

The late Georges Remi - better known as Hergé; the creator of Tintin and Captain Haddock - was reportedly planning two further Tintin stories at the time of his death in 1983. He only started work on one of them; the tantalising *Tintin and Alpha Art*, which pitted him against what would certainly have been his greatest nemesis since the probable demise of Rastapopolous in the 1968 adventure *Flight 715* which - ironically enough considering it was the subject matter of this book - featured flying saucers upon an obscure tropical island. The other postulated Tintin story which - as far as I know - he never even started to write, was to be set in the departure lounge of an international airport, for Hergé, like me, was an avid people watcher, and believed that all of life's rich tapestry could be seen in such a location. I agree with him, but would go one further. All of life's rich tapestry can be seen, if you sit quietly enough, on the great stone steps that lead up to the front of Reading railway station.

There was a courting couple, both in tears, obviously saying goodbye to each for a protracted period. There was a flustered mother looking for her lost child. There were several shady looking people in hoodies and baggy trousers who appeared to be carrying out a drug deal, and there was a young Arab man wearing richly sewn ceremonial headdress who was obviously being seen-off by his equally opulently dressed family. Watching these assorted folk carrying out their disparate lives whiled away an amusing twenty minutes for me, and before I knew it, the great double-decker coach had arrived to take us all to Heathrow Airport.

By now I really felt I was on my travels, and for the first time all week was beginning to feel quite excited. I wondered whether any of my fellow passengers on the coach were looking at me with as much interest that I had been looking at them, but for certain none of them would realise that the overweight geezer with the beard, in the front seat, was on his way to the other end of the world to search for vampires and snails shaped like flying saucers. As the coach drove along through the dreary landscape of the semi-urbanised Home Counties, I surveyed the task before me. It was going to be good to see Nick again. He and I had been friends for six or seven years, and - until he left the UK for good in 2001 - he had been a regular visitor to my home in Exeter. There had been a gap of three years before I had seen him again in Las Vegas the previous November, but it was like we had never been apart. We just picked up where we had left off, creating a slightly surreal swathe of anarchy through the otherwise ordered life of Sin City.

We arrived at Heathrow Airport just before 7.00 pm, and after an uncomfortable process of negotiating the airport shuttle (which had a facetious name which has completely slipped my mind), I ended up at a ludicrously opulent hotel for the night. The boys in Los Angeles had certainly done their job well, as my room was ready for me, there was a well-stocked bar, and a restaurant with a surprisingly nice selection of food considering its location. After settling into my room, telephoning my father to tell him of my safe arrival, and having a quick shower to rid me of what seemed to be half a ton of the local alluvium which had become attached to me during my journey, I went down to the bar to have a drink before dinner.

Hotel bars - especially those attached to international airports - are dull and uninspiring places. However, I often find that I meet someone interesting there. The night before Graham and I set sail for Puerto Rico back in 1998 we had been sat in a similar hotel bar and - to my amazement - had got talking to an elderly couple that - as it turned out - had actually known my father in Hong Kong. I sat in the corner of the bar, with my Bulldog Drummond book on the table in front me, and nursed a pint of lager. I wondered whether lightening would strike twice, and that I would once again meet someone massively interesting.

Sadly, I didn't.

I spent a dull and uninspiring evening reading 1920s pulp fiction, eating a large and expensive meal which I happily charged to the company in Los Angeles, and regretfully going to bed for a relatively early night. I decided that as I was tired there was no need to take any sleeping pills so I lay on the bed listening to the sound of the traffic outside and daydreaming about the journey ahead. I was hoping to be quickly lulled into the arms of Morpheus, but it was not to be. I spent an uncomfortable and sleepless night, and rose at about 6.00 am having given up on any hope of sleeping, had an indigestible breakfast, checked out, and made my way to Terminal 3. Despite the fact that I was there four and a half hours before the flight left, there was an enormous queue, and I spent two hours standing uncomfortably in line waiting to check my baggage in. After my unpleasant experience a few months before, I was quite expecting to have a difficult time with the security services at the airport but they smiled, waved me through, and I found that I had still several hours to kill before my flight. By this time I was really tired, and I sat around in the airport concourse trying to find something to do. Out came my trusty book, but I was so tired I could hardly focus my eyes. I wandered about aimlessly, making my way vaguely towards the departure lounge. Somewhere during this process I stupidly lost my reading glasses, which meant that for the next ten days whenever I wanted to look at a document I had to remove my glasses and squint rather than read with any semblance of dignity.

With still an hour to go, I was desperate not to fall asleep before I could board the plane, and so I ate a stodgy hamburger and drank several whisky and cokes to keep me awake (charging both to the company in Los Angeles).

Finally I boarded the plane, just after 11.00 am. It is one of the advantages of walking with a stick that most airlines will be kind and give you a bulkhead seat, which has more legroom. I was ushered to my seat by a nubile young stewardess and promptly fell asleep. When I woke up - feeling remarkably refreshed - we were three-quarters the way across the Atlantic, and I was beginning to feel that God was in his heaven and all was right with the world. Having missed my in-flight meal (which was no hardship because not only are they uniformly inedible but I had eaten far to much in an effort to stay awake at the airport), I awoke desirous of sustenance. I caught the eye of the nubile young stewardess and she brought me cheese and biscuits and a passable bottle of plonk. I ate them, decided that the in-flight movie was not to my taste, and was just about to get Bulldog Drummond out of my holdall when the bloke next to me tapped me on the shoulder, and said *"Excuse me, but aren't you Jon Downes?"*

I admitted that I was indeed and the guy (who introduced himself and for the life of me I can't remember his name) reached into his holdall, and to my amazement pulled out a copy of *Only Fools and Goatsuckers*. *"I have just been reading about you,"* he said, and I laughed, telling him that he would never believe where I was going now.

He was interested to hear that I was indeed returning to Puerto Rico, but it turned out that he was more interested in UFOs than monsters, and was convinced that the chupacabra was some kind of alien creature, which had been unleashed by its owners of a flying saucer in order to rend and tear amidst the world of men. The trouble with finding that the person sitting next to you on an international air flight is an insufferable bore, is that, unlike a cocktail party or a boozer, there is nowhere you can go to escape him. So I refrained from telling him the inadequacies within his theory, and for the two hours that it took to taxi into land at JFK we talked about the history of UFOs in Puerto Rico.

In chapter one I briefly mentioned an alleged UFO crash from 1997. This was one of a whole series of strange events, which happened at the time, which coincidentally, Puerto Rican UFOlogist, Jorgé Martin, and I investigated and found to be fallacious.

It turned out that both the stories of the UFO crash - and the marginally more interesting story of an alleged alien foetus - were both hoaxes perpetrated by the same person. Martin alleged that the hoaxer was involved in some peculiar religious cult, but did not elaborate or say whether his cultist activities had anything to do with this spreading of information.

I had originally been told about an alleged foetus from Puerto Rico by a British research Philip Mantle - a man who has fingers in many pies. He sent me a news story from Puerto Rico which alleged that a notorious undercover journalist had been found shot dead in his car outside Managua Naval Base in the south of the island. In the back of his car was found a locked briefcase, which supposedly contained a small glass test tube containing what seemed to be the foetus of an alien preserved in formaldehyde. The inference was that the journalist, whose name escapes after all these years, had been murdered by the shadowy figures of the secret world government who were trying to suppress the knowledge that they had been collaborating with alien intelligences in order to found a new world order. There was also some drivel about covert agents from the Vatican and a Nazi base in the South Pole.

In a column in the long-defunct *Sightings* magazine, I gave this farrago of nonsense short-shrift. I pointed out that there was no evidence whatsoever for secret Nazi bases at the South Pole - okay an expedition *had* been sent by Nazi High Command in 1938 to investigate the possibility of starting up a whaling station in Antarctica, but that was all. There was no link to UFOs, the Vatican, alien foetuses or anything else. Furthermore, if the journalist *had* been assassinated by shadowy figures of the new world order, hadn't they done a crappy job by leaving the foetus in the back of his car for all to see?

My dismissal of the case brought down a hail of derision on my head, but eventually I was vindicated. During the same riotous interview with the same American journalists in a Puerto Rican bar, which I have already described, I discovered that there was no mystery about the journalist's death at all. It was drug related, and the alien foetus had, in fact, been a novelty key ring. To prove his point the American journalist showed it to me. It looked like a slightly bloated version of one of those gonks that were for sale everywhere in about 1978, and was so obviously a toy, that no-one with even half a brain cell would have been taken in for more than a few minutes.

I told the story to my travelling companion but he was highly unimpressed. He had been weaned on books by people like Timothy Good that spoke of the Claymen of Saturn, Zoltan of the Fourth Density, and nonsense like that, and he was convinced that Puerto Rico was the sight of some enormous US Government cover-up.

I tried to explain that it probably was, but that the cover-up was far more prosaic than he could ever imagine. Because Puerto Rico has been denied statehood it is not covered by the same legislation that are the states of mainland USA, and therefore pharmaceutical companies and Government agencies are allowed to carry out animal experimentation and field tests of pesticides perfectly legally, although they would have been completely illegal in mainland America. Despite their legality, such experiments are carried out semi-secretly to avoid bringing down the wrath either of environmentalists across the world, or of burgeoning Puerto Rican Independence Party. He looked dumbly at me. *"Well what about the underwater bases?"* he asked. I told him that yes, whilst there had been reports of UFO like objects emerging from and disappearing into the sea, and that these reports had carried on for many years - as previously mentioned Columbus even saw a glowing fireball fall in the sea in the region, there was no sensible evidence that I was aware of that beings from outer space had established a foothold on the floor of our ocean. Moreover, as a scientist I had never seen any evidence that stood up for a moment which would support the suggestion that there were beings from outer space. We continued chatting in a desultory manner and briefly discussed some of the other UFO crashes that had been alleged to have happened on the island of the years.

He told me about the alleged incident that took place on the evening of February 19, 1984, when a UFO allegedly crashed in a slope in one of the mountains of the El Yunque National Park. The craft and several alien corpses were allegedly taken to the U.S. by military personnel working out from Roosevelt Roads Naval Station, as well as intelligence/security personnel from several federal agencies in Puerto Rico. Jorgé Martin (who claims to have been given the information by "a high ranking officer with the U.S. military in Puerto Rico", describes the incident in some detail:

"At about 1:00 A.M." - he said -, "a group of the soldiers went down in a jeep to Palmer [a sector at the foothills of El Yunque, in 65th. Infantry road] to get some cigarettes and other things in a gas station, said the officer. When they arrived to 'La Coca' waterfall sector, down in road 191, the men heard some strange noises in the brush which sounded like heavy footsteps on the dry leaves and branches in the forest. They stopped their jeep, got off it and checked the area in order to see who was there. Remember, they were there in a secret mission and no one was supposed to know about our presence in the forest. When they stopped, the jeep went dead. The lights, the jeep's engine, the radio communicator... everything went dead. Their watches wouldn't work anymore, even quartz watches."

This is another phenomenon with which I am very familiar. During our aforementioned visit to Bolam Lake, Northumberland in January 2003, when we had been investigating a series of sightings of BHM type phenomena, we had experienced a string of equipment failures. Although we had tested all of our electronic equipment the night before, charged up batteries where necessary, and put new batteries in all of the equipment which needed them, practically without exception all of our equipment failed. My laptop, for example, has a battery, which usually lasts between 20 and 35 minutes. It lasted just three minutes before conking out. Admittedly, I received an enormous number of telephone calls during our stay at the lake, but not anywhere near enough to justify the fact that I had to change handsets four times in as many hours. It seemed certain that there was some strange electromagnetic phenomenon at work here, and these electromagnetic anomalies in conjunction with fortean phenomena have been noted for decades.

According to Jorgé Martin, the men got off the jeep and took positions in the waterfall area as they kept hearing the sound of heavy footsteps in the brush approaching. It was as if someone or 'something' was coming towards them, but they couldn't see whom it was. They were ordered to be on alert and take positions in the perimeter.
He continues:

"The steps and noises continued" - said the officer - "and everyone was nervous. They shot another flare and ordered whoever it was to stop, to identify himself and to come slowly out to where they were. No answer. They repeated the order, but still no answer, nothing happens, there's no response. At this point they receive orders from their acting commander officer to shoot if necessary. That, whatever it was, kept on coming towards them. Then, the acting commander ordered them to shoot at will. One of the soldiers fired a round of shots in a crosslike motion with his automatic rifle, specially prepared for jungle combat, in the direction the sounds were coming from. Three of the bullets seemed to hit whatever was approaching.

There were more shots, the men were in very stressed. Then a lot of noise was heard coming from the forest, and the sound of something running and stepping heavily over the branches and leaves, getting away from the place. Then everything stopped and a weird calm overtook the perimeter".

According to the officer, the soldiers were ordered to a search in the perimeter. They found nothing.

Island of Paradise

Minutes later their watches, electronic equipment and the jeep started working again. They found a trail of a strange green luminescent liquid substance on the ground and leaves, following it to an undisclosed place, where they allegedly found a body. This was the first that I had ever heard about this particular incident, but I am afraid that my new companion began to appear not to believe me. I am sure that he was convinced that I was either a semi-literate moron, or actively involved in the cover-up itself, because he kept on pressing me for more and more information about various alleged UFO crashes on the island.

Sadly for him, I knew very little about them, but I *had* studied one of the alleged crashes in September 1957, and I told him what I knew about that particular incident as we taxied into land at New York.

Surprisingly, as it is the main airport for one of the biggest cities and one of the most important economic centres of the planet, JFK airport is sprawling, badly run, and in a state of disrepair. There must have been a terror alert on, because as soon as I walked off the airplane and on to American soil, the place seemed to be crawling with soldiers and armed policemen. I suppose after the events of 9/11 one cannot really blame them, but it does not do much to create a pleasant ambience for the visiting traveller, when every time he turns a corner there is a grumpy looking man in uniform brandishing a semi-automatic rifle. Despite the fact that I have read espionage novels since my teens, I am sure I would make a bloody awful spy, but I managed quite a good piece of covert operations work by nipping into the first lavatory I could find, in order to ditch my companion on the flight and his excessively dull wife, who were still talking to me at the top of their voices about how the American Government were covering up their connections with alien races. Call me paranoid if you will, but not only was I bored to tears by them, but I decided that if I wanted to avoid unwelcome attention from the authorities, to be seen to be as thick as thieves with somebody who was loudly criticising both their government and their way of life, was probably not a very wise move. After a discreet wait of five minutes, I rejoined the queue for passport control. It took what seems like a lifetime, but was in fact only about twenty minutes, for me to reach the imposing metal desk where an officer was waiting to examine my passport. He asked me a few desultory questions about what I was doing in the land of the free, and I smiled, and studiously avoided saying anything to do with American military cover-ups. Quite truthfully, I told him I had come to study snails in the caves of Puerto Rico. He looked at me as if I was mad, but quite harmless, and waved me through.

My first port of call was an ATM machine. It always amazes me that I can put the same little fag-packet sized piece of plastic into a machine anywhere in the world, punch in a number, and get out money, and I have to admit to being childishly impressed every time I do so. The world is such a smaller place than it was when I made my first forays into foreign parts, and had to buy travellers' cheques and carry large amounts of foreign currency around with me. Now I can treat each corner of a foreign field as if it was my local Post Office, and each time I do so I feel like a citizen of a global village.

It took ages to cross the airport complex to find the terminal from which I was to fly to San Juan. Twenty minutes of walking led me to a monorail terminus, and it was another ten minutes by ridiculously futuristic rail transport, before I was at my destination. The word 'monorail' has always amused me - firstly from an episode in Michael Moorcock's exquisite *Dancers at the End of Time*.[1] When the hero Jherek Carnelian meets a slightly drunken H.G. Wells at the Café Royal, he tells Wells that he is a time-traveller from the future but Wells - with the enormous ego that I can affirm is possessed by every

---

[1]. A trilogy of surreally funny books set at the end of the universe "where entropy is king, and the universe has begun collapsing upon itself". It is a comedy of manners which purports to tell the last love story of all time. Unlike much of Moorcock's work which is firmly within the canon of sword and sorcery tales, this series of books (and a few later short stories and novels set in the same 'universe' are delicately funny, and often remarkably poignant. These books come with my highest recommendation.

inky fingered scribbler - thinks that he is merely one of his more eccentric fans, and makes a few appalling jokes about how Carnelian must miss the monorails of the 21st Century. The hero stares back at him blankly not having any idea what he is talking about, and H.G. Wells goes off in a huff. There is also an amusing episode of *The Simpsons* featuring the Springfield Monorail and I sang the excessively silly song that went with it under my breath as I sped towards the terminus where I was to spend the next few hours.

> *Well, sir, there's nothing on earth*
> *Like a genuine,*
> *Bona fide,*
> *Electrified,*
> *Six-car*
> *Monorail! ...*
> *What'd I say?*
> *Monorail!*
> *What's it called?*
> *Monorail!*
> *That's right! Monorail!*

It was swelteringly hot in New York that afternoon, and as I walked the 100 yards from the monorail stop to the terminus, the air smelt of stale cooked cheese, and I was sweating profusely under the baking sun. I negotiated my way through the mass of teeming people and found a bar. There has been a dramatic change in American attitude towards stimulants over the last few decades and people, whom in 1990 were taking cocaine with their coffee, are now pointing a finger at you as a social degenerate if you enjoy a single refreshing beer. I was not particularly surprised therefore to find that there were only three of us sitting at the bar.

It is one of the social skills necessary to a good barman to strike up conversations with one's customers, and I soon found myself chatting away cheerfully with my host, and the two swarthy individuals who were sat next to me.

One was an elderly man of military bearing, in - perhaps - his late 60s, and the other a solemn looking, but wild-eyed young man about 15 years younger than me. It turned out that they were both natives of Puerto Rico, and that whereas I was flying to the island they had just flown to New York to visit relatives. The younger of the two, who turned out to be the eldest son, was called Manuel and he was an ardent environmentalist. We discussed the plight of the Puerto Rican parrot, a peculiar squat bird that was endemic to the island. *"In 1960 there were over 200 in one particular part of El Yunque"*, growled Manuel. *"But then my dad and his lot came and in 1970 there were less than twenty"*. Luckily, a captive breeding project had been set up, and within the previous couple of years several hundred individuals had been released into the wild. It looked as if the species had been saved.

But I was intrigued by what Manuel had said. It turned out that his father had been in the US Army and was involved in the controversial testing of Agent Orange on that particular part of the Puerto Rican forest. Agent Orange was one of several highly toxic defoliants that had been developed by the US military and used in Vietnam as part of Operation Hades. This had been a shameful attempt to destroy the Vietnamese rainforest and make the country into a desert where no agriculture could take place, nobody could live, and most importantly the Vietcong could not hide, from whence to carry out their extremely successful guerrilla attacks on the occupying American forces.

But it turned out that Agent Orange, disgusting as it had been, was only the tip of the iceberg. Monsanto - the company who only a few years ago caused international outrage by their irresponsible championing of genetically modified food plants - had created it. According to Manuel's father they had been tinkering with similarly dangerous biochemical technologies four decades before. Agent Orange, as it turned out was only one of several dioxin-based herbicides - there was Agent White, Agent Pink, and several others. It sounded like the cast of a Gerry Anderson TV show. They had also been tinkering with several other far more dangerous and top-secret chemical weapons.

Manuel's father didn't know any more details, but the conversation began to get ugly. Manuel and his father started to argue in very fast Spanish. I could only catch one word in ten but I did recognise Manuel calling his father a Yankee puppet, and a dealer in death. Manuel stormed off and his father turned to me. *"You must not be offended by him senor, he is Macheteros"*. I had been interested in Manuel's discussion of the Puerto Rican parrot breeding programme so I gave his father my eMail address so that he could discuss it further when I returned to England. We bade goodbye and I made my way to the departure gate.

It was only when I returned to England that I found out what *Los Macheteros* meant. It was the popular name for the Boricua Popular Army. They were a guerrilla organisation which had been founded in the 1970s and which was illegal.

For three decades they had carried out a low key but relentless war against what they saw as American oppression. Their premise seemed - on the face of it - to be reasonable enough. Some months before the Americans had won the island after the Spanish American War, Puerto Rico had achieved a certain level of autonomy from Spanish rulers. Therefore, *Los Macheteros* argued, as it was no longer a purely Spanish possession America had no right to annex it. Puerto Rico should - like Cuba, Haiti and the Dominican Republic - be an independent nation.

Although the group has claimed responsibility for numerous armed robberies and bombings since 1978, and at the time was still led by Filiberto Ojeda Ríos, a former FBI Most Wanted Fugitive, they have refocused their resources and networks to political, information, and enforcement support for the general independence and nationalist movement.

Also known as *Los Macheteros* (or 'The Machete Wielders' in English) and 'Puerto Rican Popular Army', their active membership of mostly Puerto Rican men and women have swelled to over 1100 (as of January 2005), with an unknown number of supporters, sympathisers, collaborators and informants, with cells (usually consisting of 6 to 10 members) in the United States and other countries.

Filiberto Ojeda Ríos (born in April 26, 1933) was the "Responsible General" of Boricua Popular Army, or Ejército Popular Boricua - *Los Macheteros*, a clandestine political organization based on the island of Puerto Rico, with cells throughout the United States and other countries. *Los Macheteros* campaign for and support the independence of Puerto Rico from what is characterised as United States colonial rule.

Ojeda Ríos was wanted by the FBI for his role in the 1983 Wells Fargo depot robbery in West Hartford, Connecticut as well as bond default in September of 1990. In July of 1992, Ojeda Ríos was sentenced in absentia to 55 years in prison and fined $600,000.

During the preparation of this book his luck finally ran out. On September 23, 2005, the anniversary of "el Grito de Lares" ("The Cry of Lares") members of the FBI San Juan field office surrounded a modest home in the outskirts of the town of Hormigueros, Puerto Rico, where Ojeda Ríos was believed to be hiding.

The FBI claims that it was performing surveillance of the area driven by reports that Ojeda Rios was

Puerto Rican Amazon parrot (*Amazona vittata*)

A 1969 photograph from the U.S National Archives showing a UH-1D helicopter from the 336th Aviation Company spraying defoliant on a dense jungle area in the Mekong Delta.

seen in the home. In their press release, the FBI then states they determined their surveillance team was detected, and decided to proceed with serving an arrest warrant against Ojeda Rios [1].

As the agents approached the home, shots were fired from inside and outside the house wounding an FBI agent and fatally wounding Ojeda Ríos, whom the coroner concluded bled to death over the course of several hours. No clear evidence has emerged to prove who fired the first shots.

But I knew none of this at the time, as I wandered slowly towards the departure gate.

For some unknown reason my flight was delayed for two hours, so I sat drinking Gatorade (a beverage that is unavailable in England, due to our nanny state claiming that some of the chemicals in it are carcinogenic. It is wonderful for thirst quenching, and claims to contain special agents to help with rehydration. Because of its name, I have always referred to it as 'crocodile juice', and am rather fond of it). I was feeling pleasantly drifty, so with Bulldog Drummond in one hand, and a litre bottle of Gatorade in the other, I happily wiled away the intervening time before I could embark on the final leg of my journey to the Island of Paradise.

---

[1]. Nearly all the facts and figures about the Puerto Rican freedom fighters that is given here, and in later chapters, is taken from Wikipedia. This is not just because they are a relatively small organisation, and one that even the `Terrorism` section of the CFZ library (yes, folks there *is* one) does not cover, but because there has been relatively little written about them, and what has been written that is positive to their cause was written in Spanish. Here, I felt, the idea of a consensus truth would come in useful. To test this hypothesis I checked what had been written about the Irish Republican struggle on Wikipedia, and cross checked it against what I knew to be true, and the books on the subject that were to be found in the CFZ library. They bore out pretty well, so - cross checking, using the Wikipedia references where I could - I pieced together what I believe to be a true history of the *macheteros*.

The Margarita Diary,
Chapter 3 Comments:

It never really occurred to me before I read Jon's words in Chapter 3 of *Island of Paradise* that each of us at the CFZ seem to have a character that is part-reality and partly an outgrowth, or mutation, of that same reality. Jon's description of himself as the classic Englishman abroad is perfectly accurate: that is Jon to a tee, but it is also a role Jon plays for the masses, because he enjoys doing so, and those same masses enjoy seeing it, too. And it's the same reason why I usually dress for conferences, gigs and expeditions in a black t-shirt (usually with Johnny Ramone's bowl-haired, scowling face plastered across its front) and black jeans: that's how I like to dress, but it's also slightly expected of me too. Similarly, there would be all-round shock if Richard Freeman turned up on stage dressed in anything but a frock-coat, frilly shirt, leather trousers and ornate boots.

But beneath the part-image and part-reality, the most important thing is the love for, and fascination with, all-things mysterious, and whether large or small. Hence the reason why Jon is able to get so enthused over a bunch of snails. Perhaps it's a by-product of the absurdity and eccentricity of Britain's peculiar humour, but I recall laughing – not at Jon, but with him – at the idea of us travelling to Puerto Rico under the pretext of finding one of the world's most famous monsters, just so he could get his hands on some rare snails for his burgeoning zoo back home.

As I said, however, at the CFZ we inhabit a twilight world where normality and humdrum existences play no role – and thank the great gods of cryptozoology for that. **NR**

Part Two
REVELATION

"We sail tonight for Singapore,
don't fall asleep while you're ashore
Cross your heart and hope to die
when you hear the children cry
Let marrow bone and cleaver choose
while making feet for children shoes
Through the alley, back from hell,
when you hear that steeple bell
You must say goodbye to me

Wipe him down with gasoline
'til his arms are hard and mean
From now on boys this iron boat's your home
So heave away, boys"

Tom Waits/Kathleen Brennan *Singapore*

Chapter Four
Hey Ho Let's Go...

"Dos *Medella por favor*", I said to the coffee-coloured barmaid. Until I first visited the Caribbean, I had always found the description of native people being coffee-coloured both demeaning and mildly revolting. Who, after all, would - in their right minds - liken a vision of female pulchritude to a mud coloured milky beverage of which I am not overly fond. However, coffee-coloured is the only description that really does many of the Latin American women justice. I realise that I am treading on extremely dangerous ground. My beloved (and long suffering) wife has allowed me to dictate (and dedicate) this *magnum opus* to her, and all I am doing is repaying her by describing how beautiful Puerto Rican women are.

But it is true. They are amongst the most beautiful people I have ever seen, and are not only the colour of coffee, but they move with a strange undulating grace more like a swimming snake, than any other group of human beings that I have ever met.

The barmaid smiled at me, bewitching me in a manner that guaranteed that she would get a $5 tip, and brought me a frosty glass filled with a Puerto Rican beer. Nick grinned at me. *"Get that down yer neck, Jonny boy, and I'll introduce you to the others"*.

First there was David Vassar.

I had noticed him immediately when I entered the bar. The distinguished looking middle-aged man with flowing silver collar length hair turned to greet me. We bonded immediately. It was only months later that I realised quite how distinguished the company was. David had actually won an Oscar some years before for Best Director in a documentary film. (2001, for *Spirit of Tosemite*). I liked him immediately. He was one of those educated, slightly alternative blokes who sadly you meet more and more rarely as the world becomes a brasher and less-alternative place. Within minutes of meeting, we discovered we had many interests and tastes in common, and as we began to discuss the project that had brought us to Puerto Rico I soon decided that we were going to be firm friends. Then there was Kevin, the cameraman. Cameramen are a funny breed, being both highly skilled technicians and artists, and one often finds that not only is the cameraman the most highly paid member of a film crew but also that he has probably led the most interesting life as well. On my previous trip to the island Nick Plowright - a veteran of Everest expeditions, South American insurgencies and a hundred other things besides - had filmed me. Even now I regularly see his name on the credits of things I am watching on television, and can quite justifiably say with pride that I know that man. Kevin was a quiet, slightly shy and immensely affable, dude. We shared similar senses of humour and liked most of the same music. Over the next few days, whenever the subject of vampire hunting palled, one would be sure to find me and Kevin sharing a stupid joke, or talking about the music of Pete Townshend.

All of a sudden, the door to the bar flew open and a coffee coloured whirlwind burst in *"hell and goddammit, those stupid son-of-a-bitch bastards of the airport sent me to the wrong terminus"*, she said as she came over to join us. It was Carola, a Puerto Rican resident who had been hired by the company in Los Angeles to be producer, general factotum, and all-round Ms. Fixit for the expedition. A girl of about thirty, she was plumply attractive, but had a wicked sparkle in her black crystal eyes. Within minutes every man in the bar - and certainly every man in the crew - fancied her something rotten. She was perfectly aware of this fact, and used it to lazily manipulate us into doing her bidding for the rest of the trip.

She handed out itineraries to us all, and we discussed the programme for the next few days. I was happy to hear that the next day we would not be filming but would be exploring the island, looking for locations, and general acclimatising. Carola turned to me. *"So Jonny, your luggage is - how they say - up sheeet creek?"* I nodded, swallowing a smile at her eccentric use of the English language. She said, *"Tell me the things you want and I shall have them for you by the morning."* As practically everything, apart from my medication, my camera, and my collecting equipment were in the interstices of the American Airways bureaucratic machine, I gave her a list of essentials, and thanked my lucky stars that having been caught out like this before I always have a spare pair of jeans and t-shirt in my hand luggage.

Much later the party broke up and Nick showed me my room. The residential part of the hotel was actually a couple of blocks away, and the two of us, carrying on our friendship as if there had never been an interruption, walked slowly through the darkened streets bathed in silver moonlight, listening to the sounds of *el cocqui* and the insects of the night towards an unprepossessing apartment block. For security reasons we had to open a massive wrought-iron gate which led into a small courtyard. Nick ushered me through, locked the gate again, and we found my room. We said our goodbyes; I showered, took my medicine, did a brief inventory of what possessions I had left and fell into a dreamless sleep.

The next morning I awoke feeling happier and healthier than I had in years. I always turn the air conditioning on full when I am in the tropics, and I leave it on all day to make sure that the room is nice and cold for my return.

Until relatively recently scientists wondered how cold-blooded reptiles such as the leatherback turtle were able to survive in cold northern seas well inside the Arctic Circle. They discovered that it was only the hugest marine reptiles that were able to do so, due to a process called gigantothermy. Basically, the larger an animal is, the smaller its surface weight ratio is and the longer it takes to retain or lose heat. As animals get larger they have less surface area - skin area- relative to their volume, so they loose less heat to their surroundings. Think of a cup of tea and a bath of water at the same temperature. The tea will cool relatively quickly as it has a large surface area: volume ratio, but the bath with its smaller surface area: volume ratio will keep warm for longer.

Being a big bloke, I suffer from this quite badly, and whilst it has its advantages; in the winter I can wander around in a short sleeved shirt while other people are tucked up in their winter woollies, in the sum-

Kevin (striped shirt) and David

Nick and Carola - even the most feeble-minded reader will not need telling which is which

ABOVE: Boat tailed grackle *(Quiscalus major)*
BELOW: One of several species of *anolis* that inhabit the island

ABOVE: Our home from home BELOW: One of the many species of cocqui (stolen from a Hawaiian newspaper, because although I have heard them often enough, I have still not managed to *see* one)

mer it sometimes takes me so long to cool down, that sometimes I don't get to sleep until the dawn is breaking in the east. This problem is even greater in the tropics, and so I make sure that in the interests of obtaining a reasonably early night, that my room is in the Palaearctic while all about me swelters.

It was only about 8.30 am when I left the apartment complex, and in a genteel, sedate manner I ambled down to the main part of the Windchimes Hotel for breakfast. As is my habit in the tropics, I had my breakfast outside. *"dos tostados por favor"*, I said, asking for one of the nicest breakfasts I have ever eaten. They are like a sweet Danish pastry filled with cream cheese, and I have never eaten them anywhere else in the world. My breakfast in Puerto Rico is always the same. A pint of freshly squeezed orange juice, two of these delicious pastries, and a couple of slices of papaya washed down by delicious Puerto Rican coffee. I was just finishing and preparing to light my breakfast cigarette, when the rest of the crew ambled into the bar looking slightly worse for wear. Despite the fact that most of my luggage was God knows where, I looked reasonably spruce that day where I suspect that David and Nick at least, had mild hangovers. We sat around at the table smoking and talking, until just before 10.00 Carola ambled in with a broad smile, and carrying a bulging carrier bag which contained a care parcel for me. I took it with me, as we walked out to the two Japanese people-carriers that were waiting for us.

David gave me a walkie-talkie, and gestured that Nick and I should travel in the car with Carola, whereas he and Kevin would follow in the smaller vehicle. The next few hours were a wonderfully evocative trip down memory lane for me. Apart from Carola who lives there, I was the only one of the crew who had spent any length of time on Puerto Rico, and I was certainly the only one who had scouted the area for film locations. I think I earned some brownie points with Carola, and certainly with David, when I suggested over the walkie-talkie that as we were entering old San Juan we should turn right down the *Calle del Sol* towards the main plaza, where there was some stunning colonial era architecture which might well make a suitable backdrop one of the pieces to camera which we were to film over the next few days.

As we drove along, I renewed my friendship with two of my favourite Puerto Rican inhabitants; the boat-tailed grackle *(Quiscalus major)* - a strange jet-black bird which looked like a mentally-deficient starling with an extraordinary rudder-shaped tail - and the small anole lizards *(Anolis spp.)* - (I have no idea which one for there are no less than eighteen species on the island), which scuttled up and down every available wall sunning themselves, and carrying out a lizard-orientated lifestyle. We stopped at the plaza that I had recommended, and whilst David and Kevin discussed camera angles, I introduced Nick to these two eminently Puerto Rican animals. Delighted, we watched as a large male anole and a member of the grackle tribe, who seemed even more retarded than his peers, argue over the remains of a breakfast burrito. The male lizard, puffing out his chest and brandishing his brightly coloured throat flaps, raised himself up to his entire height of four and a half inches, and faced-down his opponent. The grackle may have been many times his size, but was obviously wiser than he looked. The tiny, but ferocious, lizard attacked the black bird's feet, and chased it away, and then laboriously dragged the remains of the burrito towards a hiding place in a nearby bush where he could eat it in comfort.

It was time to move on, and we drove down the narrow streets passed avenues of gloriously baroque buildings, each painted in contrasting and very bright colours. There is something carnival-like about old San Juan. There is music in the air - the strains of the local salsa, come out of every shop and bar window, and it is no exaggeration to say that as you drive past somewhere where the music is particularly strident you see the local people literally dancing in the streets. This is part of the reason why I love Puerto Rico so much. It has a joyous, life-affirming property about it that I have experienced seldom anywhere else in the world. Despite its bloodthirsty history, it is a place where the best of Taino, Spanish and West African culture has met, melding and forming something utterly unique and very, very, precious.

As it was getting towards lunchtime, we decided to head out of the city towards the town of Canóvanas where most of our filming would take place. Over drinks the previous night I had mentioned to Carola how much I enjoyed the local delicacy of *mofongo*. Everyone's ears pricked up; it is something, which I have noticed amongst film crews all over the world that they like to try the local food. I explained happily, that *mofongo* is one of the nicest things I have ever eaten. It is basically mashed plantain with garlic and butter. It can be served like a mashed potato or it can be made into cricket ball-sized lumps and deep-fried. It is the Puerto Rican equivalent of the chip butty, and whilst you will not find it in any of the *cordon bleu* restaurants, every roadside café sells it. Carola told me that she knew one such café on the road to Canóvanas, which *"made the best mofongo in the island"*, and she promised to take us there for lunch. My mouth was watering in anticipation, as we drove along the surprisingly deserted road. As soon as we left the urban sprawl of San Juan (the elegant old town was now surrounded by extremely ugly and modern conurbation whose lack of architectural taste would have made our present Prince of Wales have apoplexy), the countryside became wilder. The buildings became more tumbledown, and were usually surrounded by small patches of scrubland. Every couple of miles there was a tiny patch of jungle, maybe thirty yards wide, and I gazed rapturously out of the window, and counted the different species of wild birds.

After about forty-five minutes journey, we pulled off towards the side of the road, and parked outside a dilapidated looking building which had once been quite an attractively built restaurant and now looked like the sort of thing you expect new age travellers to erect at one of the less well-moderated British rock festivals.

When I was a child in Hong Kong we spent most weekends on our boat, the *Ailsa*. My father co-owned her with another family in the colonial administration, and - until she was sadly destroyed in a typhoon in 1970 - we used her most weekends to explore the little islands, which make up the archipelago of Hong Kong. People think that Hong Kong is just an area of urban sprawl, but in fact it is made up of over 300 islands, some of which are uninhabited and some of which, at least back in the 1960s - when the world was a younger and more innocent place - only had little fishing villages on them. Each weekend in the *Ailsa*, and later in a friend of my father's cabin cruiser, we would visit one of these islands. I never knew which I preferred; the uninhabited ones where I could stalk the local wildlife, and occasionally find such exciting things as a fox hole which had once been the lonely hide out of a Japanese soldier awaiting the American invasion, or the inhabited ones where there were other children to play with, and usually a corrugated iron shack on the beach which sold cold drinks and ice cream. Some of the happiest times of my childhood were spent of these beaches, and as we walked towards the dilapidated restaurant I had a surprisingly vivid flashback to the days when I was a schoolboy.

David sent me and Nick to a covered area at the side where underneath a makeshift canopy, which appeared to have been constructed of canvas and brightly coloured bunting, we sat at a large round table surrounded by granite seats. Nick disappeared to answer a call of nature, and came back a few minutes lately, grinning broadly, sweat dripping off his shaved head and brandishing two large cans of ice cold Medella. *"Life don't get no better than this loike"*, he grinned at me.

"You haven't tried mofongo yet", I said.

A few minutes later, the rest of the party joined us. Then a procession of small boys wearing threadbare clothes, and barefoot, appeared on the scene bringing trays of food to our table. I had expected *mofongo* but what I got was one of the most amazing meals I have ever eaten. It was a veritable smorgasbord of different Puerto Rican delicacies. As well as the *mofongo* there were balls of rice and garlic, plates of barbecued pork, grilled chicken, strange exotic seafood and baked manioc tubers.

The roadside café and the mofongo feast

You can take the boy out of the West Midlands but you can never quite take the West Midlands out of the boy

Nick looked at me a bit old fashioned like.

"Er, we don't have anything like this in Walsall loike", he said uncertainly, and I decided, not for the last time that week, that you can take the boy out of the West Midlands but you can never quite take the West Midlands out of the boy. *"For God's sake Nick, you are not still pining for the Rotunda are you?"* and we began to joke in cod West Midlands' accents. Our companions looked at us with confused expressions. What they didn't know is that Nick and I always pretend to have arguments over his cultural origins. It has become a running joke, and we do it almost without thinking. Americans who have never heard of the Birmingham Bullring (The Bull Ring market has been an important feature of Birmingham since the Middle Ages. The market began in the year 1154 when Peter de Birmingham a local landowner, obtained a royal charter. Initially a cattle and food market, it developed into the main retail market area for Birmingham as the town grew into a modern industrial city), who have no idea that the Rotunda is an extraordinary coke can shaped building in the centre of Birmingham, and think that Birmingham is in Alabama, really haven't got a hope in hell of understanding our peculiar sense of humour.

I'm afraid I started it all off when I wrote my novel, *The Blackdown Mystery* in 1999. Although based on real events (basically most of what I said happened did happen, though not necessarily to the people that I said that it happened to, or in the order that I put in the book), it was fictionalised enough to be a novel, and I had amused myself by depicting Nicky R as a West Midlands oik obsessed by punk rock music. OK, that is not too far from the truth, but he is also a sensitive, intelligent, and very nice man who is also one of my best friends.

I should have been prepared though, for a few years later Nick lampooned me back in his book *Three Men Seeking Monsters* in which he described me as a somewhat effeminate alcoholic. Since then I have been threatening to get my revenge, and taking the Mickey out of the West Midlands whenever I get the chance is part of the process. I described Nick in one magazine article as "pining for the Rotunda", and claimed that his Dallas home is a shrine to the West Midlands, and that he insists on having mushy peas, Tizer, and H P sauce flown out to him on a regular basis, together with crates of Special Brew…

It is all light-hearted nonsense, and almost completely untrue, but I am the first to admit that it is probably completely incomprehensible to those people who are around us, and not in on the joke!

After our magnificent meal, we got back in the cars and drove towards the national rainforest - usually called El Yunque after the most well known mountain in the range - which is situated in the highlands above Canóvanas. The journey took the best part of an hour, but it was an excessively pleasant way to spend an afternoon. The mountain was as beautiful as I had remembered and, as I had done on my previous visit, I marvelled at how quickly the landscaped changed from uninspiring scrub to the tall grassland plateau above Canóvanas, and then again to lush rainforest as we drove up towards the summit. For the first time I had a chance to look at the grassland plateau. I knew of its existence of course, but for one reason or another during our first visit to the island, our paths had not strayed that way.

I was impressed with David. Having discovered that the vast majority of chupacabra sightings, if not attacks, have taken place in this plateau, he insisted that we explore it thoroughly.

Once again I was reminded of my boyhood in Hong Kong.

Whereas once upon a time the whole of the island of Hong Kong had been lush jungle, it had been deforested at various times during the past few centuries, and although the jungle was growing back, there were areas where the main vegetation was either bamboo or tall lush grass similar to the pampas grass which you can buy in any garden centre, but far less refined with a wiry feral quality. It was here that I had played as a child, creeping through the tall stems which towered five or six foot over the ground, and living out my boyhood fantasies of Red Indians, Robinson Crusoe, and assorted Swallows and/or Amazons. During our weekends, when we visited some of the smaller islands, which were too small to have ever been able to support a forest environment, I again explored these weird grassland areas, and marvelled at the rich eco-system of small animals that lived there. There were beetles, lizards, snakes (although I never told my mother, who was always terrified that I would get bitten) and on several occasions I even found the places where one of the most spectacular of Hong Kong mammals, the Himalayan porcupine (*Hystrix hodgsoni*) had been. The only times I ever saw live porcupines were occasionally at night when my father was driving us home after dark. Once or twice I remember seeing the distinctive black and white shapes lumbering at the side of the road. However, I saw them dead as road-kill several times, and once during an expedition, which was so secret I have never told anyone to this day, I found one of their skulls.

I had been visiting some family friends at a place called Mount Kellett. Like the apartment block in which I had lived, it had been built to house senior Government officials, but unlike my domicile at Peak Mansions, the countryside immediately around Mount Kellett was much more exciting to the naturalist. Peak Mansions was an exciting enough place to be - within a 100 yards of our front door was the top of a long winding footpath which led through a couple of miles of deep forest down to the reservoirs at Pokfulam, and a little further on the Lugard Road - which circled Victoria Peak and from whence one could see snakes and lizards aplenty and a variety of the local birdlife. But Mount Kellett boasted a far lusher stretch of jungle, and one could regularly see barking deer, civet cats and - so it was rumoured - the remnants of the once plentiful population of dhole or red dogs. It also boasted one of the best resources for children that I have ever seen. Successive generations of colonial sprogs had built a massive tree-house complex in two or three interlocking banyan trees that grew, gnarled like old men with arthritis, only a few yards from the politely manicured lawns of the apartment block. I loved to visit Mount Kellett - I had several friends there, I always enjoyed playing in the tree-house and at the age of ten - although my days of actively pursuing the fairer sex were well over half a decade in the future - I was beginning to evince some curiosity about the female of my species, and, for some reason, the free and easy atmosphere which seemed to prevail amongst the youthful population of Mount Kellett was conducive to my being able to satisfy my curiosity.

One afternoon, while my mother was drinking tea and chatting to her friend, I and the daughter of the house, went to explore. We climbed over the small perimeter wall that bordered the Mount Kellett car park and found ourselves within only a few minutes deep in unexplored jungle. Whatever thoughts I had had of a game of "Doctors and Nurses" disappeared, as whatever genetic predisposition I have of being an intrepid explorer took hold, and for the first, but certainly not the last time in my life, I put exploration and pursuit of knowledge before my interest in the opposite sex. Stealthily the two of us explored and found, much to our surprise, that there were a couple of small, dilapidated, but obviously still inhabited huts deep in the jungle only a couple of yards away from the block of flats which we knew so well. This was amazing! At the time I spun a complex story to explain them. Even then I was interested in cryptozoology and had also read a number of pieces of classic juvenilia in which heroes like Biggles discovered lost tribes. OK, with hindsight, I think even at the age of ten I should have realised that a lost tribe of primordial Chinese would not be found only a couple of yards away from a block of flats in which resided dozens of senior civil servants, but at the time I thought I had made a momentous discovery. The huts were - also we found out a few days later - where the elderly Chinese couple who maintained the gardens of Mount Kellett resided. They had been built only twenty years or so before by the Urban Services Department, and there was no mystery whatsoever. However, we didn't know this as we poked and prodded our way around, and found that just downhill from the huts was a small but flourishing kitchen garden. At the side of this garden were two wire traps, one containing bones and quills, and the other containing the very putrefied corpse of a dead porcupine.

I carefully opened the trap, removing the skull and some of the quills. I still have them somewhere as a *memento mori* of the first of many times in my life where I eschewed women in favour of adventure.

Back in 2004, as we drove through the high grassy plateau with occasionally stands of bamboo, it was easy to slip back into memories of my colonial childhood. We drove past the occasional house, but unlike the shanties further down the hill, these were obviously prosperous dwellings. They were well kept - usually of two storeys with a large circular veranda wrapped around them - and festooned with satellite dishes. Carola told us that this was a sought after neighbourhood. Government officials and rich businessmen lived there. The weather was better; it was cooler and less noisy. Also evocative of my childhood were the flame trees, which were just coming into flower. I believe that flame trees originated in East Africa but I am not sure. Wherever I have been in the world, however, they seem synonymous with a benevolent colonial administration. I am not sure why but wherever the British Empire stretched in the tropics and sub-tropics, every high ranking Englishman abroad decided to plant one in his garden, and it seems that they were equally popular amongst the Puerto Rican counterparts.

Every so often we saw a farmer. However, unlike the tumbledown shanty farms I had visited years before in the lowland districts, these were much more prosperous affairs. The road dipped slightly as we drove into a valley, which, apart from the flame trees and tall pampas grass, could have come straight from my native Devonshire. On the right hand side there was a farm gate, which could have been found on the outskirts of Bideford or Barnstaple, and Carola told us that this was going to be one of the main locations for filming in a few day's time. Suddenly something extraordinary happened. A small group of brown hairy creatures only eighteen inches to two foot in height lurched unsteadily on to the road in front of us. Carola slammed her foot on the brake, and we shuddered to a halt. They were such peculiar looking animals, and in most un-mammalian behaviour they seemed to move en masse like a shoal of fish or a flock of birds, that it took me a few instants to realise what they were. They were goats.

I made some facile remark to Carola, and she told me that this was the most common type of goat kept on the island. They were like pygmy goats that you see in a petting zoo at some down at heel seaside holiday resort in the UK, rather than the sort of animal that comes to mind when one usually considers a goat kept as a domestic animal.

We drove on and soon found ourselves at the bottom of the rainforest. By now it was early afternoon and we knew we had about four hours left before sundown. We started looking for locations in earnest. I had been particularly interested in the fact that many chupacabra sightings seem to take place along the banks of rivers, and so I suggested to David that it would be a good idea to do some pieces to camera by one of the mountain streams which cascade like silver snakes through the dark green verdant foliage which carpets the craggy mountains like an ill-fitting wig. We examined several, and - to my pleasure - I had a chance to go and explore. If I had known I would have brought my net with me. I had brought a fishing net with me in my collecting equipment but, like an idiot, had left it behind at the hotel. However, happy as Larry, I clambered over the moss-covered boulders and, making sure not to break the delicate heart's tongue ferns, which shrouded them, I found myself re-living another favourite pastime of my childhood - investigating natural streams.

My mother used to say that there was no point in dressing me up neatly if we were going anywhere near water. I was bound to go off and investigate, probably fall in, and come back with my pockets stretched out of shape with various jam jars full of wiggly things. This time I had no jam jars, nor did I fall in, but I got satisfactorily muddy as I poked around in one of the little pools, which had been formed by the swift flowing mountain stream. There is nothing quite like them in the UK; what mountains we have are usually quite bare and desolate, and although there *are* streams, and indeed streams which hold nice little eco-systems of their own, there is nothing that compares with a mountain stream in a tropical jungle.

The tiny mountain stream and the author preparing to go and do his own aquatic thing

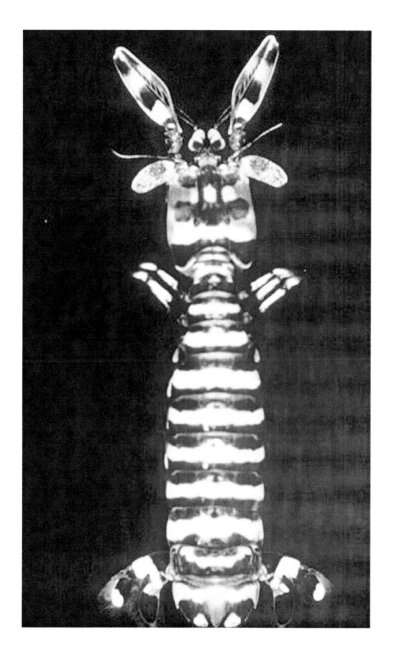
Mantis shrimp

Island of Paradise

Just sitting on the giant boulder, I could see a myriad of life forms. On a rock opposite me there was a tiny skink, and a bird that looked like a wagtail of some description was standing on the shore just upstream of me singing his heart out. In the water were tiny corydoras catfish, and some little minnow-like fish, which I was unable to identify. The pool was only perhaps four foot across but I estimated that it held at least forty fish. There were little multi-coloured freshwater crabs, and - always being interested in such things - I debated whether to try and capture some and take them back alive to my collection in Exeter. I decided not to; I had just about enough equipment to bring back a viable quantity of cave snails, and I thought it likely that the little crabs would perish before I got them home. I clambered downstream to the next pool about thirty feet below, and there I saw some extraordinary shrimps.

Mantis shrimps are so called because of their feeding strategy. Like the well-known praying mantis they keep their two front claws bunched up in a reverential posture beneath their neck, from which they can leap out, punch and grab their victims.

Called "sea locusts" by ancient Assyrians, and now sometimes referred to as "thumb splitters" by modern divers - because of the relative ease the creature has in mutilating small appendages - mantis shrimp sport powerful claws, formed like jack-knives that they use to attack prey.

The species are commonly separated into two distinct groups determined by the manner of claws they possess: "spearers" are armed with spiny appendages topped with barbed tips, used to stab and snag prey and have a blunt, calcified club on the elbow, while "smashers" possess a much more developed club and a more rudimentary spear; the club is used to bludgeon and smash their meals apart. Both types strike by rapidly unfolding and swinging their raptorial claws at the prey, and are capable of inflicting serious damage on victims significantly greater in size than themselves. These two weapons are employed with blinding quickness, rapidly reaching 10 m/s from a standing start, and can strike with a force comparable to a small-calibre bullet.

Some mantis shrimp, which are sometimes kept as aquarium pets, have managed to break through their double-paned aquarium glass with a single strike from the weapon. Smashers use this ability to attack and feast on snails, crabs, molluscs and rock oysters; their blunt clubs enabling them to crack the shells of their prey into pieces. Spearers, on the other hand, prefer the meat of softer animals, like fish, which their barbed claws can more easily slice and snag.

Mantis shrimp appear in a variety of colours, from rather pedestrian browns to stunning neon. Their eyes - both branching from a single stalk - are similarly variably coloured, and are considered to be the most complex eyes in the animal kingdom. Each eye possesses trinocular vision, and some species have at least eight different visual pigments sensitive to various wavelengths, and three more sensitive to ultra-violet light. By comparison, humans have only three visual pigments. Mantis shrimp also have four filters that tune those visual pigments, they see two or three planes of polarised light, and each eye is capable of depth perception independently of the other eye. [1]

These ones were a light biscuity brown, but they were undoubtedly mantis shrimps.

I was shocked away from my reverie by David and Nick who were calling me to return to the car, and continue scouting the jungle in search of locations for our film.

We drove up the hill further into the heart of the jungle. On my first trip in 1998, I had been disap-

---

1. The information on mantis shrimps is culled from a variety of sources, not the least being www.blueboard.com/mantis/ The Lurker's guide to Mantis Shrimps, Simon Wolstencroft pers. Comm. and - of course - Wikipedia.

pointed by the paucity of wildlife, and I drew tentative conclusions that the low number of invertebrates - in particular - that we saw was something to do with over use of untested pesticides on the lowland farms. That may well have been the case, but I was beginning to think that there must just have been something wrong with that particular part of the forest being visited, because in this part of the jungle I counted a dozen species of butterflies and several other types of less beautiful insect. I even saw a Puerto Rican parrot - out of the corner of my eye - as we sped up the hill. The further we went, the lusher and more impenetrable the jungle about us seemed to be, thus it was even more of a surprise when we turned a corner and were confronted - on the left hand side of the road - by a large lay-by featuring the tackiest tourist attraction that I have seen outside Blackpool. When you looked closer, you realised that it had a bizarre kitsch charm all of its own. Everything was handmade for one thing, and coming from a culture where "handmade" usually means that it has been made by one of a select group of artists to be sold at a hideously inflated price to members of the cognoscenti – it was somewhat of a rude awakening to be reminded that in a peasant economy (and this may have been a tacky tourist trap, but it was a tacky tourist trap owned and run by peasants who were largely living hand to mouth), "handmade" merely meant that the people who made it could not afford machines.

There was a café and a gift shop, but the things that really caught my attention were two enormous garishly painted sheets of plywood with brightly painted forest scenes upon them. Like something from a seaside town in Britain thirty years ago, they each contained stylised comedy sketches of people - Tarzan and Jane, and a bathing beauty and a Charles Atlas type - with holes cut out so that visitors could pose having their photograph taken. To my immense amusement one of them had the legend "El Chupacabra" painted about the gaps where the heads should go, and Nick Redfern and I could not resist being photographed posing there.

Over lunch I had confided in both David and Carola of my desire to catch some Puerto Rican cave snails. I suggested that we revisit the caves at Aguas Buenas and try and get some decent film of the cave carvings, which I had seen there on my previous visit. As well as the carving of the snail there was also - what looked like an incredible ancient - rendition of the *sheela na gig*, a human fertility symbol which is known from many cultures across the world. The name is Celtic, and describes a female figure sitting cross-legged with her reproductive organs heavily stylised into sharp relief. The example that I had seen in the Aguas Buenas caves also had a stylised image of the sun on her head. It was obviously being used for ritual magickal purposes still, because there were fresh candles as well as small offerings of flowers for it. Back in the dark days of 1998, we didn't have a digital camera, and - in the best tradition of Fortean evidence everywhere - the photographs that I had taken in the cave had not come out.

Sadly, it looked as if we were going to be unable to go back there on this trip, and although David and Carola both liked the idea of filming in some caves, I was disappointed to find that I was quite likely not going to be able to collect my precious snails after all.

After Nick and I had posed, looking ridiculous, which is now on the front page of our website, we stopped off for a cold drink. As we were walking up to the café I was startled to find a snail - looking very much like the ones I had found in the cave so many years before - sitting minding its own business on a banana leaf. With a deft swoop of my wrist, I plucked it and put it straight into one of the Tupperware containers, which I had brought for the purpose. I showed it to Carola who said, *"Oh I know those things, they are everywhere!"* And she was right. I collected half a dozen in the next few minutes, and at two other locations later than day I collected more, hoping that I would have a varied enough gene pool to produce a viable breeding colony. But there was one thing that was puzzling me. These snails were undoubtedly herbivorous. Over the next few days I provided them with a wide range of plant life, and they gladly ate them all. A year later they were still happily munching away on cucumbers and dandelion; two plants that they would never have encountered in Puerto Rico. But the two snails which I had

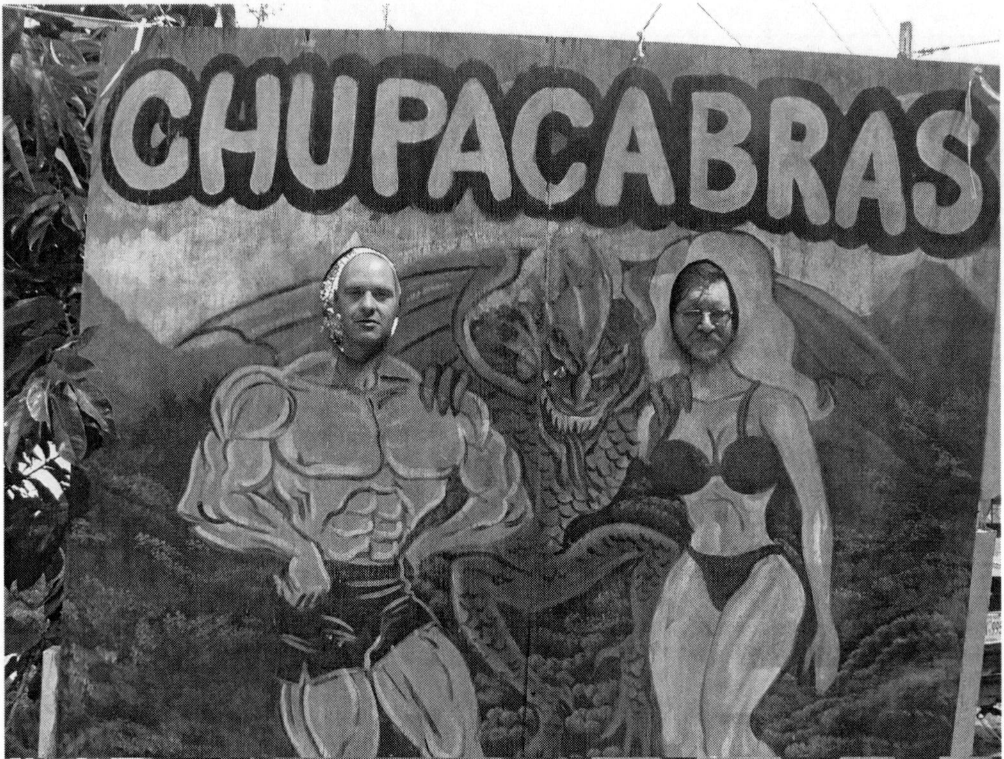
CHUPACABRAS

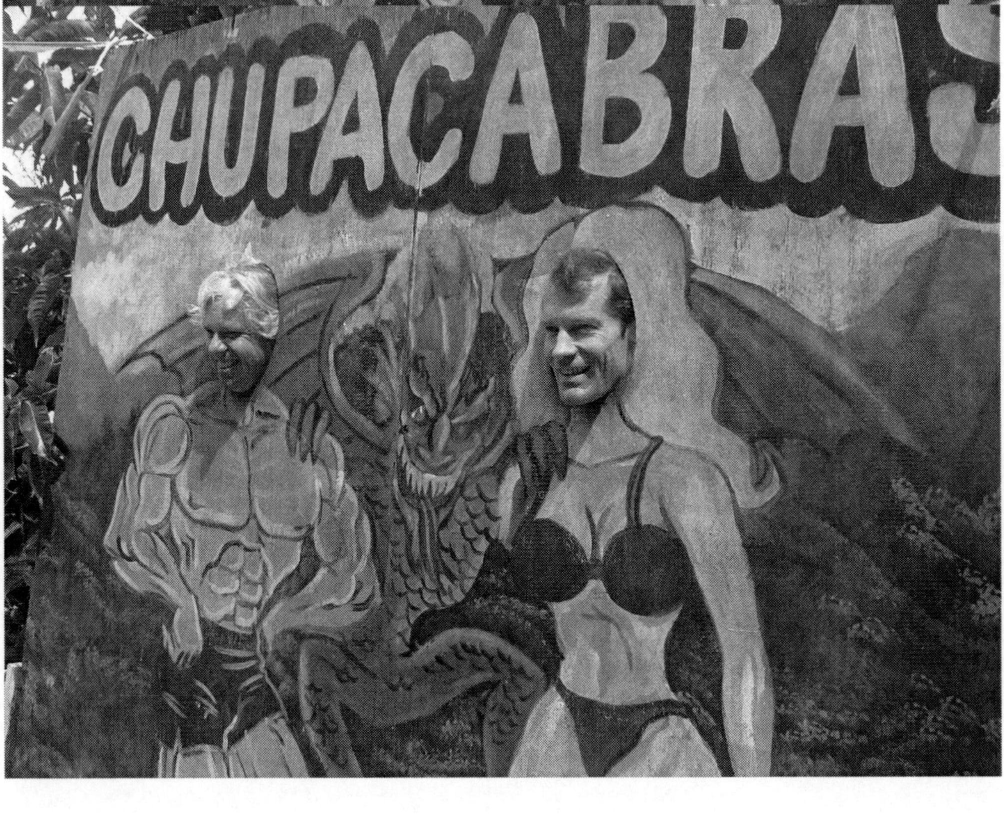
CHUPACABRA

Carola with coconut BELOW: Forest snail in natural habitat
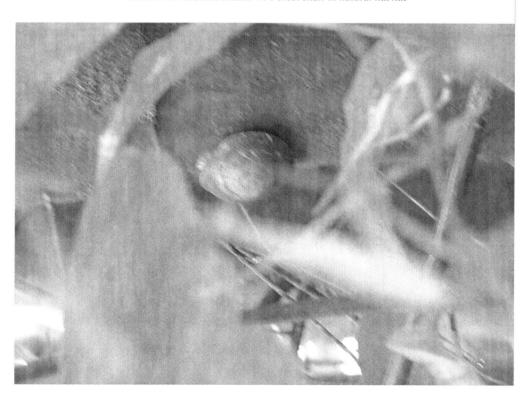

The strangest snails that I have ever seen. I later identified them as *Nenia tridens*

The author with one of his precious snails

El Yunque raimforest

caught six years before had been subtly different. The shells of my new acquisitions were green with lichen and moss, and their bodies were a healthy dark brown colour. The snails I found in the caves - which by the way had fed off bat droppings and mushrooms - were paler in colour and (from memory) a little larger. Could they be a different species? Or was this just an incredible adaptable species, which would eat anything and live anywhere? I was determined to find out, and decided that the only way that I could prove either viewpoint was to carry on with my original plan and attempt to set up a breeding colony back in Exeter.

However, I was still hoping that we would still manage to go to some cave systems, where I could find some true cave snails, and so I decided to keep my spare collecting jar empty in anticipation of such an occurrence. I have been kicking myself ever since, because only about thirty feet away I saw three of the strangest snails that I have ever seen. I later identified them as *Nenia tridens*, but they were so peculiar in shape - looking more like caterpillars than anything else - not only did I long to have some for my collection, but I felt certain that they would command high prices on the exotic invertebrate market in the UK. If I am ever lucky enough to return to Puerto Rico, I am going to make certain that I take more collecting equipment along. Upon my return to the UK when I started investigating the molluscs of the island I discovered that there are a number of highly specialised and peculiar snails, as well as at least one species of velvet worm (a highly primitive animal which has remained almost unchanged since it and its kind were amongst the first creatures to colonise the land hundreds of millions of years ago) living on the island, and - to the best of my knowledge - none have ever been kept in Europe.

Nick and I bought bottles of cold drinks, and walked back to Carola's car. I was amused to see that someone - presumably her - had written with her finger in the ingrained dust on the back window. In curly, and very feminine, handwriting it read, *"I am the Chupacabra, eat me"*. This made no sense whatsoever, but only endeared Carola to us more. We had only spent a few hours with her, but she had already wormed her way into our affections. She was a mine of information - and misinformation - on the subject of the island and its culture, she knew everything, and she knew everyone. Wherever we went, whenever we got out of the car, someone would flash a beaming smile and wave and shout, *"Hey Carola!"* It was obvious that they were all very fond of her, and treated her a bit like a younger sister who was always getting into mischief, and who you never really knew what she was going to do next.

Having refreshed ourselves, we drove further up the mountain until the road came to an abrupt stop. We got out of the cars when we could go no further and found ourselves in a veritable sound storm of *el cocqui*. I had only ever heard one or two at a time before, but here there must have been hundreds.

Even a single one of these tiny frogs makes a weird whistling, shrieking sound, but to hear them en masse was truly breathtaking. They are undoubtedly the unofficial emblem of Puerto Rico, and a great deal of folklore has built up around them. One story says that they were originally birds who had somehow transgressed and been turned into frogs by God Almighty. Taking pity on them however, he had given them the power to climb trees again but not to fly. Another story says that they are the descendants of a mischievous child who was turned into a frog for another set of unknown transgressions.

One of the biggest pieces of folklore about them is a contemporary one. It is said that they will pine away and die if taken away from their beloved Puerto Rico. This is - sadly - not true because not only are they also found on various of the Virgin Islands, but within the last few years, some were transported to Hawaii in a shipment of tropical plants and have since multiplied. Hawaii is a very fragile and isolated island eco-system and we have already seen what damage a tree snake from Indonesia has caused. It is certain that these noisy little frogs will have some affect on the balance of nature there but presently it is too early to tell. Even during this trip to the islands I was again told how they could not be taken away from Puerto Rico, and I hadn't the heart to enlighten my new friends.

Carola told me that nobody knows exactly many species of *cocqui* there are. Some estimates are that there are as many as eighteen - each subtly different in the intonation of its diagnostic call, and it has been claimed that two species have become extinct within the last thirty years. Once I could hear large numbers of these tiny frogs singing their hearts out, and I could tell that the calls were indeed, subtly different, and I think that this is yet another excuse - if an excuse were needed - for me to take a party from the Centre of Fortean Zoology back to Puerto Rico for a third, or even a fourth, investigative trip.

It was now beginning to get late, and we still had business to complete back in San Juan. We turned the cars around and slowly made our way down the hill. On the way back we identified four or five different locations where filming could take place over the next few days, and all of us were in a very upbeat mood. On the way down the mountain we stopped off again at the tourist attraction. There was an observation tower there and David and Kevin wanted to see how viable it would be to take the camera to the top of the tower and get a panoramic view of the forest and valley below. I sat in the car and watched my new snails, as Carola wandered off for a few minutes, returning with a semi-ripe coconut which someone had hacked a few holes in with a machete and inserted a straw. She grinned at me gaily, *"Here you try it, it is really good."* I did, and it was. And we spent a massively entertaining return journey down the mountain; Carola driving at breakneck speed, Nick in the back seat making dreadful jokes, and me in the front. Carola and I passing the coconut between us until it was dry.

We drove back towards San Juan. For the first time I took real notice of the environmental deprivation in the region. After having seen the lush beauty of the mountain forest, to drive through what felt like miles upon miles of ill-kempt building site was depressing. I said as much to Carola and she shrugged her shoulders, *"People must eat,"* she said, *"and these building projects bring work."*

Soon we were entering the outskirts of San Juan. I had been so busy bird watching, snail hunting, shrimp spotting, and taking photographs of anything and everything, that I could that I hadn't really been paying attention to the conversation that had been going on desultorily between Nick, David and Kevin all day, so I was actually quite surprised when we pulled off the highway, and instead of parking up outside our hotel I found that we were in a seriously dodgy looking used car lot. David leapt out of his car with immense vim and vigour and said, *"OK boys, we are going to get you a jeep"*.

Carola went with David, and Nick and I stayed in the car. He was as enthusiastic as me about the day's events, but like me he was looking forward to a hot shower, a cold beer, and an early night. Then Carola came back to the car, and asked Nick - who is a registered resident of the United States - to come with her. As he has a US driving licence, it made more sense for the jeep to be hired in his name, but what neither of us were prepared for was the long tortuous process of picking out a jeep, establishing our *bona fides*, and actually hiring the damn thing.

It took the best part of half an hour, and I sat in the front seat of Carola's car, bored at first - but then entranced to see a cat-sized animal which looked like a strangely misshapen rat stalking gingerly up and down on the other side of the perimeter fence. I recognised it as an opossum, but in the half-light would not be able to tell you the species. Puerto Rico has a very depleted mammal population. There are only thirteen native species known to exist on the island and all of them are bats. Several others, including two species of ground sloth, and some medium sized highly specialised rodents are known to have died out before, or soon after, the Spanish arrived.

The other mammals on the island are all introduced. The Indians brought dogs with them and probably guinea pigs for food. The Spanish brought rats and pigs, and several other mammals had been introduced by accident or design over the years. The rats had forced the Puerto Rican shrew, (*Nesophontes edithae*) into extinction, and quite possibly done the same to other species as well. I had heard rumours

Island of Paradise

that chimpanzees and rhesus monkeys had escaped from laboratories in the 1950s and 1960s and had become naturalised in the west of the island, for a while at least, and in the late 19th Century some bright spark had decided to introduce mongooses to the island. Mongooses are exclusively old world carnivores and are found across Africa and Asia, one species even ventured into Europe. The species that had been introduced to Puerto Rico was the great Indian mongoose and for many years it was thought that this invasive species had caused the extinction of the Puerto Rican nighthawk, which had not been seen since 1961, but which had recently been rediscovered. How or when the opossum had been introduced I did not know. But they were certainly there - I had just seen one!

I began to get depressed as I always do when I reflect on what a Godawful mess my species has made of the planet on which we live, but was shaken out of it by the arrival back on the scene of the rest of the party.

"Jonny boy, you are going to love this," said Nick and he showed me a smart looking jeep. Grabbing my holdall containing my precious snails, I followed him to it. *"Now Puerto Rico will see that the boys from the CFZ have really arrived"*, said Nick with a grin and we got in. Shouting out goodbyes to Carola and telling David and Kevin that we would see them in the bar, Nick switched on the engine and drove out of the compound. *"I have been wanting to do this all day"*, he said, and plugged his portable CD player into the car radio slot. He pressed the on button and the unmistakably sounds of *The Ramones* playing *Blitzkreig Bop* blared through the speakers at full volume.

> *"Hey ho, let's go-*
> *They're forming in a straight line*
> *They're going through a tight wind*
> *The kids are losing their minds*
> *The Blitzkrieg Bop*
> *They're piling in the back seat*
> *They're generating steam heat*
> *Pulsating to the back beat*
> *The Blitzkrieg Bop.*
> *Hey ho, let's go*
> *Shoot'em in the back now*
> *What they want, I don't know*
> *They're all revved up and ready to go."*

Truly at times like these it was good to be alive!

The Margarita Diary,
Chapter 4 Comments:

I'm pleased to see that Jon devoted so much space to a discussion and description of our time spent at the *Windchimes* hotel. For those who weren't there, it might seem superfluous; but for Jon and me it was a time to rekindle a friendship that had been separated by the Atlantic for a couple of years; and it was a time to make new friendships with David, Kevin, Carola, and the rest of the crew.

There is something unique about the camaraderie that comes with hanging out alongside fellow thrill-seekers and adventurers – all from different corners of the globe, most not even knowing each other, yet all thrust into a strange and surreal quest to seek out the truth about a diabolical beast said to roam a real-life paradise.

I can safely say that those first few days as we all got to know each other, shared jokes, experiences and tales over breakfast, lunch and late-night dinners about our respective cultures and lives were both enlightening and joyous. It was a time to celebrate and toast life, and to do whatever the hell all of us wanted to do before our collective clocks stopped ticking.

But it was without doubt the day we got our hands on that shining, silver jeep that things really took off... NR

Chapter Five
Call me Ishmael

Now we had a vehicle of our own, and were thus, at least *semi*-autonomous, Nick and I felt that the adventure had started for real. The next day we were up bright and early, and had finished our breakfast, and were milling around in the bar with expectant looks on our faces when David and Kevin, and a few minutes later Carola, arrived.

Here I would like to say a few words in self-mitigation, before the folk, who seem to have made it their life's work to criticise me every time I even *mention* alcohol, raise their ugly heads. The reason that we congregated in the bar each day, was that it was the only part of the hotel complex which actually backed on to the road. The bar also doubled as a breakfast room, and was a pleasant and convivial place in which to meet, and we met to work and not necessarily to drink beer. In certain quarters I have been accused of allowing my memoirs to become set mostly in bars and saloons. There is a certain amount of truth in these accusations, but in this case, at least, I am innocent....

When Carola came in, a stocky man with very short, cropped, steel-grey hair accompanied her. He was the sound engineer with whom we would be working for the next few days, and she introduced him to us as "Cheese". For the next two days I was convinced that this was a diminutive of 'Jesus', (usually pronounced "Hey-Soo"), which is not only the given name of our Saviour but is a popular name across much of the world. But no. His name was indeed "Cheese". Apparently, when he was younger, he had a shock of bright red hair, which was extremely unusual on the island. His friends had said that he looked like an Edam "Cheese" and the nickname had stuck. "Cheese" was a lovely man, and did much to open my eyes to the rich and infinitely variable Puerto Rican culture. Nick and I introduced ourselves, and "Cheese" wired us for sound.

This is quite an intrusive process which involves shoving the transmitter of the radio microphone either in your trouser pocket or down the front of your pants, and threading a wire leading to a transducer microphone up the inside of your shirt, and sticking it to your chest with adhesive tape. It is a very useful tool to film making, but has a real disadvantage; for so long as you have the thing on it is impossible to have a private conversation, and one has to be very guarded with one's tongue just in case someone is listening.

We spent the first few hours driving around old San Juan, past the fort, where in the 16th Century, Sir Francis Drake had tried unsuccessfully to invade the island, past the undeniably imposing Parliament buildings, and round and round the maze of tiny terracotta streets with their tall brightly coloured colonial era buildings. This was the route we had driven the day before, but this time we were being filmed; first from behind, then from the front and finally with "Cheese" and Ernesto, his assistant, perched in the back of the jeep. Nick and I were given the difficult task of extemporising a conversation about our plans for the forthcoming week, and in fact, we were in the amusing situation of having to film several different versions, depending on what we might find when we started our investigations for real.

Then we drove through the tiny streets of old San Juan along the seafront, through the new and ugly city, and out into the countryside towards our first port of call.

For a variety of reasons we had not been able to find the true location where six years earlier I had taken the feather samples, but Carola had managed to find us a farm which raised roosters for cock-fighting, and which had, indeed, only a few years previously suffered the predations of the goat sucker. For me, having visited a farm like this many years before, it was no great shock, but I believe from the look of dismay in his eyes that Nick felt quite upset at visiting such a squalid location and seeing serried rows of wire cages, each containing a solitary cockerel doomed to die in the arena. What made things worse, was that some of the birds were in an appalling condition and a few of the cages contained dead or dying fowl. The owner; a shifty looking fellow with bloodshot eyes, and a pock-marked face did not impress me. He told us how, in 1997 or 1998 (he couldn't be sure), he had come to his farm one morning to find that every single bird had been slaughtered. He took us to the place where his birds were kept. We had to negotiate several high wire fences of the sort you have around tennis courts, each only traversable by a locked metal gate. We walked across a concrete yard in which streams of liquid slurry ran from a shippen at one end. Over the barn doors we could see a couple of scabby and unhappy looking cows, and wandering around myopically were two or three of the tiny Puerto Rican goats, each looking more dejected and badly cared for than the last.

I was convinced that this level of security would be impossible for any animal of the size of the chupacabra to pass, and asked the farmer - whose name was Ramon - whether the fences had been up at the time of the incident. I was fully expecting him to say that they hadn't been, and that they had only been erected to keep the chupacabra out, but he grimaced at me in what I think was meant to be a smile, and answered *"Si Senor, they have been here since my father's time"*. We finally reached the coops with the fighting cocks, and Ramon told us how, on the morning in question, he had found that the wire had been twisted, but that the bolts on the cage doors had actually been opened. This was a weird one. To the best of my knowledge only a man, or at the very least a higher primate with an opposable thumb, would have been able to open it. He told us how each of the animals had been found dead, drained of blood with puncture marks in the neck, and that a number had had the top of their heads bitten off, or with triangular holes in the skull through which the brain had been eaten.

"Were there any footprints?" I asked, but was told that there weren't, but he had found some strange coarse hairs stuck in the wire of one of the cages. They were not human, nor horse, nor cat, nor anything else he could identify. He had given them to a policeman who had investigated the killings, but had never heard anything else about the matter.

Having conducted our interview with him, we then had to film it, together with establishing shots showing us arriving and leaving. This is where we ran into problems. Late the previous night, while Nick and I were having dinner in the barn, the funny little moonfaced man who had welcomed me to the *Windchimes* when I had first arrived, came and told me that my luggage had finally caught up with me. I was glad to hear this and so during our first days filming I was resplendent in tropical white jacket, Panama

The picturesque streets of Old San Juan

ABOVE: The parliament building in San Juan BELOW: The fort, where in the 16th Century, Sir Francis Drake had tried unsuccessfully to invade the island.

ABOVE: Cheese BELOW: Ismael

Filming at the smallholding where the chupacabra attack had taken place
L-R David, Ramon, Carola, Kevin

Island of Paradise

hat and rather natty trousers. I soon found that the climate was so hot that I could not wear them, especially the jacket for more than a few minutes without them being soaked through with sweat. I had recently been diagnosed as being diabetic and possibly as a result of this, and certainly as a result of being quite a few stone overweight, I tended to sweat more than more people. In this climate it was so uncomfortable to be almost unbearable; at least if I was going to wear my respectable clothes. It was OK in shorts and a t-shirt, which I had worn the previous day, like 99% of the male population of the island, but if I were to maintain my 'Our Man in Havana' image, then a certain amount of artifice would be necessary. In the gaps between takes I took my jacket and hat off, and allowed my body to breathe. During the journeys from one location to the next, I changed back into t-shirt and sometimes shorts and drank copious amounts of water. It got so bad that by the time we reached the zenith of heat at midday, Paolo - the young son of one of the crew - was deputised to follow me around carrying an icebox full of litre bottles of mineral water.

We finished filming at the farm, bade our farewells and drove away. As we did so I noticed the slogan *'Morigua Liberdad'* painted on the wall of one of the farm buildings. I asked "Cheese" what it meant.

He told me that Morigua was the original name of the island before the Spanish had invaded, and that, whereas *de facto* independence campaigners were actually complaining about American rule and not Spanish rule, despite the fact that they were of Spanish descent, they had adopted the ancient Indian name for the island as part of their battle cry. *Morigua Liberdad* was one of the battle slogans of a group of extremists, which I later found out were the *Macheteros* like Manuel, the young environmentalist that I had met in New York.

As we drove along I suddenly realised after about twenty minutes that I had no idea where we were going. My white jacket and Panama Hat, together with my collecting bag were laid out relatively neatly in the tiny space behind the seats where, presumably, one's children, one's shopping or one's trained dwarf was supposed to reside. *The Ramones* were blaring out of the speakers as Nick drove along singing happily and blithely out of tune in a broad Walsall accent.

> *"Chewin' out the rythym on moi bubble gum*
> *The sun is out and oi want some*
> *Its not hard, not far to reach,*
> *we can hitch a roide to rockaway beach"*

I reached back and grabbed the walkie-talkie. Managing to make myself heard above Nick, Dee Dee's and Joey's guttural vocals, I asked David where we going and what we were planning to do next. In fact, I said *"Earth to Vassar, Earth to Vassar, whats'a happening dude?"*, whereupon the radio crackled and David gave a creditable impersonation of Adam West at his mid-60s best, and said and almost shouted, *"To the bat cave, boys"*.

With remarkable sharpness of mind, I answered *"Holy Puerto Rican microchiroptera, Boss. Roger, over and out"*, sat back, took a swig out of yet another bottle of iced water, and continued a completely fatuous conversation with Nick Redfern about mutual acquaintances back in good old Blighty.

Soon we were driving along a coastal road, and, having dropped enormous hints to Carola that some local seafood would go down a treat for lunch, preferably washed down with a teensy weensy glass of the local plonk, I looked hopefully out of the window for a bijou little seafood restaurant with a nice wine list and an ocean view. Sadly, there was nothing of the sort, but after another few miles we pulled in to an unattractive piece of scrubland. Switching off the car engine but leaving the air conditioning running, we waited to receive instructions. They were not long in coming. The radio crackled and David's voice sounding for all the world like an enthusiastic young scoutmaster trying to raise morale amongst a particularly recalcitrant band of boy scouts, said: *"Come on boys, we are here"*.

Here? Where? I couldn't see anything of even the slightest interest. Carola got out of her car looking extremely pleased with herself. *"It is great isn't it?"* she gushed. Was I missing the point? Was I suffering from a touch of the sun? Or was I just being particularly dense? I must have looked a bit of all three, because Carola, looking quite concerned, came over, patted me on the shoulder with such intimately conspiratorial terms that one would have thought she was making an indecent proposal to me *"It is a cave. Come and see my lovely cave"*.

I still couldn't see anything. All I could see was an unprepossessing cliff face on the other side of the road, a few scabby looking thorn bushes in front of it and a steady stream of little birds flying to and forth. I walked across the road and saw to my complete embarrassment that I had been fooled by a trick of the light.

About fifteen years ago, everybody I knew seemed to have an obsession with those damn fool *Magic Eye* books. No matter how hard I looked I could never make any discernable pattern emerge, and I am afraid it was the same now. I had been staring straight in the mouth of a shallow but substantial cave. The little birds, which appeared to be some kind of swift, were flying in and out and as I got closer I could see that amongst the stream of birdlets there were a fair number of small bats. The midday sun had been so bright that it had illuminated most of the interior of the cave and made it appear flat and two-dimensional when in fact it was no such thing. I went in, and one of the crew who had been following on with "Cheese" in an articulated lorry shouted to me *"Be careful Senor, those are - how do you say this - bloody t'ing... vampire bats"*.

I had been told before that there were vampires on the island of Puerto Rico. Indeed, Graham and I had filmed some bats high in the roof of the cave system at Aguas Buenas and two zoologists had independently identified them to me as being vampires. Strangely, though, according to the official literature, although there are thirteen species of bats known to the island, none of them are, or have ever been vampires. There can be only two explanations. One - and I am sure that this is the explanation that most readers of this book will prefer to be the case - is that the fauna of the island is not so nearly as well known as the authorities believe. There would certainly be plenty of precedence for this. The annals of mainstream, yet alone cryptozoology is full of accounts of animals turning up in places where they are not supposed to be. One particularly good example of this was the family of foxes, which turned up on the Isle of Man in the late 1980s, hundreds of years after it had been conclusively stated that foxes never inhabited the tiny island.

The second, and sadly more prosaic explanation, is that the presence of vampire bats was just a native superstition. This conveniently ignores the evidence of the people who saw Graham's and my video but in my defence, I would say that the video footage is not of particularly good quality; the camerawork is shaky and the resolution poor.

However, one of the things that I have learned over the years is that when one is a guest in someone else's country, one has to behave with respect.

So when one of the crew - to whom I had not actually been introduced at that time - shouted *"Be careful of the bats senor!"* at me as I entered the cave, I waved at him cheerfully, and promised that I would be.

The heat hit me like a steamroller as soon as I entered the cave. In all my years knocking about the planet, I don't think that I have ever been quite so hot. It was a searing humid heat that made walking

The author in the batcave

dreadfully uncomfortable, but I was determined to search for some cave snails, so braving the impossible heat, I began to search the cave to no avail.

The cave wasn't really a cave at all. It was a concave dip in an almost sheer cliff face, maybe sixty foot high and thirty foot deep. I wish I had paid more attention to my physics lessons at school, so I could have explained why it was so excruciatingly hot in there. However, I didn't, so I can't.

I examined all the parts of the cave wall that I could get to, and found no snails - indeed no invertebrate life whatsoever. The only animals that seemed to have made their homes in this inhospitable environment were the tiny bats - which the crew still insisted were vampires, and which I must admit *looked* like vampires - and the tiny swift-like birds which had made their nests alongside the hanging bats high in the domed roof like that of a guano stained cathedral high above. As I searched for snails, the sweat was pouring off my face, it trickled into my eyes and I found myself mopping my brow with a handkerchief, and later a towel, more than I was actually searching for snails. However, I was desperate to try and obtain some of these weird beasts, and I continued my search.

Just as I had finished, David called me over. It was time for me to perform a portentous piece to camera about the history of the chupacabra phenomenon. I basically told the camera a very much briefer version of what is in chapter one of this book and, as always seems to be the case, David had to film bits of my monologue from different angles for cut-away shots. The only real problem is continuity. It was so hot that I couldn't have my jacket on for more than a few minutes without it being soaked through with sweat, and despite the fact that I was swigging down iced water at a remarkable rate (I drank something like seven litres in under an hour), I was finding it hard to stand up for too long during the ten or fifteen minute breaks between shots. Young Ernesto came to the rescue (prompted by Cheese), and bought be an orange crate to sit on between takes and continually refreshed my water bottle. Luckily, I had brought several changes of shirt with me. Months before I had been shopping for clothes in an enormous mall in Terre Haute, Indiana with my friend Jessica. I had been pleased to find half a dozen almost identical Van Heusen shirts for sale at $5 a time. I bought them all. She asked why I had bought so many shirts that looked almost exactly the same, and I told her when making a TV documentary, sometimes bits are edited together which sometimes took place several days apart, and thus to avoid continuity problems it helps if you always wear the same apparel. In the documentary that was made of our first trip to the island, there is an appalling continuity gaffe which no-one apart from me and Graham seems to have noticed. You see a long shot of Graham and me walking along a forest road in the hills above Aguas Buenas, and it cuts to us beginning to climb a mountainside. Unfortunately, I am wearing a denim jacket and t-shirt in one shot and a blue and white lumberjack shirt in the next. I did, and was determined that this would never happen again.

That day in the seaside bat cave I was very grateful for my forethought, as I actually went through four shirts in a period of half an hour. I was glad when my stint of filming was over. I dragged my orange box, moved out of shot and sat down to watch Nick take his turn. I am glad that I did, because if I had not I wouldn't have witnessed on of the most amusing incidents of the whole trip, and furthermore one which is perfectly true, and which goes some little way towards my exacting some revenge for Nick's depiction of me in *Three Men Seeking Monsters*.

During my first trip to the island, in the caves above Aguas Buenas, Graham - who had been filming the bats high above us in the cave roof - complained that he felt something strange trickling down the back of his shirt collar. I was delighted to inform him that he was probably the first cryptozoologist in memory to have a vampire bat pee down the back of his neck. It was delightfully reminiscent of the Monty Python sketch when Oscar Wilde, James Whistler, and Bernard Shaw are making up more and more nonsensical adulations to the future Edward VII.

Island of Paradise

OSCAR: Your Majesty is like a big jam doughnut with cream on the top.
PRINCE: I beg your pardon?
OSCAR: Um... it was one of Whistler's.
WHISTLER: I never said that.
OSCAR: You did, James, you did.
(The PRINCE OF WALES stares expectantly at WHISTLER)
WHISTLER: ... Well, You Highness, what I meant was that, like a doughnut, um, your arrival gives us pleasure... and your departure only makes us hungry for more.
(Laughter)
Your Highness, you are also like a stream of bat's piss.
PRINCE: What?!?
WHISTLER: It was one of Wilde's. One of Wilde's.
OSCAR: It sodding was not! It was Shaw!
SHAW: I... I merely meant, Your Majesty, that you shine out like a shaft of gold when all around is dark.

However, all of this was knocked into a top hat by what happened next. Not only did a bat, which may or may not have been a vampire, high in the rocks above his head, urinate on to the top of his shaven dome, but because of the lack of hair it splashed everywhere, making delightful prismatic rainbow effects in the beams of sunshine which filtered in through the undergrowth in front of the cave. I burst out laughing and ruined the take, mildly to David's annoyance. When I explained what had happened, the whole crew fell about laughing, and I felt that I had somewhat of a moral victory.

After another half hour or so, during which we filmed establishing shots of me and Nick walking into the cave and walking out again, we drove up the road and found somewhere for lunch.

In the bar on our first evening I had expressed the wish that whilst in Puerto Rico I wouldn't do anything that I was able to do at home, and would always eat legitimate Puerto Rican cuisine. Carola had promised to do what she could, and so far had been as good as her word, but for some reason we needed to pay a visit to a shopping mall on the outskirts of the nearest town, and whilst David and Carola did whatever they had to do which I think from memory involved a bank, we sat and ate at a depressingly corporate American fast-food joint with an uninspiring salad bar. I was quite surprisingly hungry after our morning's exertions and ate heartily but it was an unappetising, and not very satisfying meal, and so I wasn't at all disappointed when we left and made our way to Canóvanas.

Although it had been six years since my previous visit to the town, I found to my surprise that I recognised quite a lot of it. Back in 1998 we had spent a frustrating day waiting around for the legendary Mayor Soto - the first politician to take the chupacabra attacks seriously - and his lieutenant, Ismael Aguyo, from the town's civil defence department. We never did meet the Mayor, and when, after something like six hours, Senor Aguyo turned up I don't think either of us were particularly impressed with each other. I said as much in *Only Fools and Goatsuckers* and I realise now that I had taken the man completely wrong. I did not know that we were actually going to meet him again, and had only been told that we were going to rendezvous with a local expert. We drove through the town, and I was pleased when we drove past the gates of the Civil Defence depot without entering. However, in a little lay-by one hundred years down the road we pulled in and I instantly recognised Ismael as he got out of his car to meet us.

"Oh bloody hell", I muttered to Nick. *"This doesn't look promising"*. But Ismael came rushing up to us with a broad grin, patted me on the shoulder with a semi-hug, and shook my hand *"Mi amigo, it is good to again see you"*, he said in faltering English, and he really looked like he meant it.

Over the next few days I discovered - to my eternal embarrassment - what had gone wrong. As already mentioned, 1998 was the height of chupacabra fever and I hadn't realised quite how many European and American film crews had come to Canóvanas merely in order to make fun of its inhabitants. Whereas our documentary had not done anything of the sort, indeed it was me that had been made fun of rather than anyone else, Ismael and his crew had understandably treated us with caution. Now, six years later he was glad to see that I was back. It meant that I was serious about my quest, and once he found that I was a scientist who treated the chupacabra phenomenon as very real and interesting, he was determined to do all in his power to help us.

He went over to Nick Redfern and I introduced them. To my great joy he muttered *"Call me Ismael"*, totally unaware of the fact that he was echoing Herman Melville from a hundred and something years before. I have always been an avid fan of *Moby Dick*, probably because one of my mentors, Tony 'Doc' Shiels, loved the original Gregory Peck movie, and I am always overjoyed when I can slip a little reference of it into my day-to-day life. I saw Ismael's unconscious pandering to my wishful film fantasy as a good omen, and was convinced that this time we were going to get closer than ever before to solving some of the islands enduring mysteries.

Our little convoy was now five vehicles strong. In the lead was Ismael, closely followed by David and Kevin in their car. Then came Carola's people carrier, followed by us in the jeep (still blaring out the raucous sounds of 1976 at full volume), and making up the rear - a heck of a long way in the rear - was a large articulated lorry containing the crew, the sound gear, the lighting gear, and presumably quite a lot else besides. I never quite understood why such a large articulated lorry was necessary. Although "Cheese" and his boys used quite a lot of equipment, it could have all been fitted into the boot of a fairly large car, but no doubt they had their reasons, and wherever we went the large lorry followed us about twenty minutes behind. Our little convoy drove up into the mountains above Canóvanas through the high grassland plateau and up to another series of hills.

"I don't know why", I said to Nick, *"but this place looks familiar"*.

We drove on, and subtly the country began to change. We were near the top of another small mountain - nowhere near as high as El Yunque. It was not covered in rainforest, but was draped in a mixture of thick scrub and occasional other trees. The further we climbed, we could see that it was one of the foothills of El Yunque itself, but I kept on seeing pieces of the landscape that I found familiar. We reached the crest of the hill, me still not knowing why or when I had been here before, and Nick gently teasing me that I was having some strange psychic *déjà vu* experience. Then I remembered. The last time I had been on this stretch of road it had been in an enormous civil defence truck, sirens wailing, emergency lights flashing, and me, Graham and a gaggle of swarthy worthies holding on for dear life in the back, while Nick, the cameraman, was lashed to the roof of the cab.

"It's the place with the weird fish", I said to Nick, and much to my gratification he immediately knew what I meant. I told him the story of how we had visited this little valley for the first time, filming under particularly peculiar circumstances. This had been the occasion when I had attempted to teach some of the Canóvanas National Guard to sing *The Rocky Road to Dublin*, and where I found the first zoological anomaly which had led me on the road to my still unfounded theory, which is basically the reason why I am writing this book.

When I was a little boy in Hong Kong, together with many of the other youngsters who lived at Mount Austen Mansions - an apartment block where I resided between 1964 and 1968 - I spent many of my leisure hours at a place called Tadpole Pond. Although it was full of tadpoles, it wasn't actually a pond; it was a concrete basin, which had been built during the war years by the occupying Japanese, presuma-

ABOVE: The stream BELOW: *Gambusia affinis*

bly for some arcane use involving motor vehicles driving up and down the Peak Road. However, now it lay derelict, and was only ever visited by the youth of the neighbourhood who went there to catch fish, red freshwater crabs, and tadpoles.

It was full of tiny silver grey fish, which we called guppies, but which were actually gambusia – or mosquitofish. They are a species native to the southern states of America, which in the early years of the 20th Century were introduced across the tropics and sub-tropics of the world in an attempt to control malaria. Although not native to Puerto Rico, they had been introduced there in the 1920s, and I was pleased, when on my first visit to this little valley - which we were just entering - to find a stream, which contained a thriving population of them. Graham and I had caught one and taken a photograph of it, but it was not until many years later, when preparing slides for my presentation at the UFO Conference in Las Vegas, I looked at the picture again and noticed some peculiar morphological differences between the fish we had caught, and a normal specimen of *Gambusia affinis*. This had triggered a series of incredible thoughts in my mind, and I was overjoyed that we were able to go back there and that I might be able to test my hypothesis.

The mosquitofish is a small, stout fish with a small head and a terminal, strongly upturned mouth. It has a small, rounded dorsal fin that originates behind the anal fin. Dorsal fin rays are seven in number, and anal rays are usually the same. The first few rays of the anal fin are greatly elongated in adult males. Mosquitofish do not lay eggs, but rather give birth to live young. These fish, therefore, require no special environment, as most other fish do, for depositing and hatching their eggs. They breed throughout the summer, and new broods are produced at intervals of about six weeks, with 50 to 100 young in a single brood. The young are approximately $1/4$ inch in length when born and grow to a maximum size of about three inches. They eat a diet of aquatic and terrestrial insects, microcrustaceans, and other invertebrates.

We found ourselves parked up in a strange little valley. Although I had been there before, Nick hadn't and I think that he, like me, was quite surprised by it. There were flowering trees everywhere, and steep, jungle hillsides leading up to what would eventually become the summit of El Yunque. At the bottom of the valley were a couple of large, and quite well-appointed houses, a small flock of chickens and two or three of the ubiquitous Puerto Rican goats wandering around. It seemed prosperous and well tended, the birds sang and only the occasional cockcrow shook a jarring note. By the smaller of the two houses there was a small growth of bushes and out of the corner of our eyes we could see two or three small children - wearing the minimum of clothes necessary to preserve their decency - hiding in the bushes whispering to each other and pointing at us. They were obviously excited to have visitors, but too bashful to come out and see what we were doing.

I knew why we had been brought here. There was a cave, and a weirdly unearthly heap of giant boulders positioned so one could clamber through the gaps, halfway up the hillside. It had been a difficult climb up there six years before, and I knew from experience that there was just not enough light to film satisfactorily, and although the chupacabra had been reported to be reputedly living there a few years previously, but the caves and the boulders were accessible, it would have taken me the best part of an hour and a half to get up there, and as the enormous articulated truck thundered over the rise of the hill, and gingerly negotiated the flock of chickens to come to rest next to our cars, I told David that, in my opinion, it would take several hours to get me and all the equipment up there, and that having seen the location some years before it just wasn't going to be worth it.

David was inclined to agree with me, especially as we only had a few hours of light left, but he decided - being a consummate professional - to send Kevin up with Ismael to carry out a reconnoitre, and to see whether my suggestions would prove to be correct.

This suited me down to the ground. I had a completely different agenda, and was overjoyed to be back in this little valley. Although, as I have already said, my plans for the Puerto Rican expedition included several pieces of mainstream natural history that I needed to investigate in order to flesh out my burgeoning theories, I never for a moment thought I would ever be back in this valley, and I was thrilled to be there.

Until recently I was the deputy editor of one of Britain's leading tropical fish magazines. I know a fair amount about the subject, but it so happens that *Gambusia affinis* - an otherwise unremarkable species - is one of the fish I know most about. It is important in the history of tropical fishkeeping in that it was the first tropical fish species ever to be kept in the home aquarium. It is so hardy it can be kept under practically any conditions, and some eighty years ago when the hobby was in its infancy, and the only way tropical tanks could be heated was by having paraffin burners positioned under the sloped bottoms of the tanks, which - as you can imagine - caused some extreme fluctuations in temperature and water chemistry, this was the only fish species hardy enough to survive. It was this hardiness that allowed it to be transported all over the world for the purposes of malaria control, but it was also this hardiness, which led - along with the fact that it is pugnacious and cannot be kept with other species in an enclosed tank - that led to its downfall within the hobby. Unlike other more attractive and less hardy species, although it can be kept under practically any living conditions, it doesn't mutate or show colour morphs. With the exception of a close relative - *Gambusia holbrooki* - it does not interbreed, even with the other members of the gambusia tribe, and as such, whatever you do to it, it remains a slightly dull, dark grey bullet-shaped fish.

It was the first tropical fish species that I managed to breed, and I have studied them both in captivity and in the wild. In November 2003 when Graham and I were preparing my photographs to illustrate our trip to Puerto Rico I noticed something very strange about the photograph, which he had taken. Rather than having the diagnostic pointed nose of most gambusia, this fish was snub-nosed - like a bulldog. I wanted to know exactly what they were, but until - totally by accident - we had returned to this stream, I had no way of knowing: Was it a fluke? Or were there other snub-nosed individuals in the stream?

Nick and I strode over to the stream. Much of it was covered with sheets of corrugated iron, so we carefully lifted them off and to my relief I saw that the stream had not become polluted, and was still the home to a large and thriving shoal of gambusia. To my great joy, when I looked closely I could see that within this shoal there was a sub-group - maybe 20% - of fish which were undoubtedly gambusia apart from having this strange snub-nose. However, although the stream was only a few inches deep the fish were wary, and as soon as I dipped my net into the water they disappeared. I tried again after an interval of half an hour but to no avail. Eventually I caught one and could confirm, yes it had a strange snub-nose. I was forced to let it go because I had made no preparations for such an eventuality. I had no test-tubes, no formalin and no way of preserving a dead specimen, no way of humanely procuring such a specimen, and certainly no way of successfully transporting a living one back to Britain.

I showed it to Nick, but until when - later in the expedition - I explained the enormity of what we had found, I think that he merely put my excitement down to one of his strange friend's eccentricities.

Nearly a year after the event I started writing the main body of this book, and I was sitting in the house of my long-suffering girlfriend (now my wife) in Lincolnshire. She foolhardily offered to type out this book as I dictated it, and I am immeasurably grateful to her for having done so. Despite the fact that it was late May, the effects of global warming were making themselves felt; that afternoon there was a hail storm and thunder, followed by rain of almost tropical intensity. I found it very difficult to share with her, let alone put on paper; the great feeling of almost *religious* joy that I experienced on finding out that it appeared that my theory about the gambusia was true. That there were indeed a distinct population of bizarrely misshapen fish in the tiny mountain stream.

But I was experiencing an almost Pauline sense of revelation, and furthermore, one which is almost impossible for a writer of limited talents, like me, to express in mere words.

What I wanted to know then, and was beginning to understand, was what had caused this mutation. Although species mutate all the time, most mutations are not viable and result in stillbirths or infertile eggs. When a mutation is viable it has to be of some use to the species or some benefit in order to survive. We are not just talking about one or two of these little snub-nosed fish but an entire population. Although I was beginning to realise what could have caused the mutation, the important thing that I now have to work out was how such a modification could be of any use to the species as a whole.

Suddenly I realised that my disappointment in not being able to capture some of my cave snails that I had seen so many years before was unfounded. If I was correct, the forest snails, which were by now happily eating their way through mountains of vegetation in a pink Tupperware sandwich box in my *en-suite* bathroom back in my hotel room in San Juan, would do perfectly for my needs.

Lost in thought, Nick and I wandered back towards the cars.

Then Kevin and Ismael rejoined us. Kevin confirmed that the cave was not really suitable for our purposes especially as it was beginning to rain and the light would be gone in another hour. I breathed a sigh of relief. It had been a difficult climb, made worse by the fact that the hillside was infested with strange poisonous plants, including one particularly virulent herb, which had an array of psychedelic bright pink berries. It looked like a much larger version of the British cuckoo pint, with one great difference; even if you brushed lightly against the berries with exposed skin, you came up in violent blood blisters within sixty seconds. They had been so painful, and the reaction so severe, that I am sure that if a person with a less cast iron central nervous system than mine had encountered one of these, it could well have killed him, or at the very least sent him into anaphylactic shock. We all got back into our respective vehicles and prepared to leave.

We drove in convoy once again back down the hill towards Canóvanas. Now I hoped we would have a chance to check out another one of my own personal objectives to this mission. Six years before I had caught and held a *bona fide* unknown animal. I wanted to try and find the place where I had caught it, examine the location, photograph it in detail, and - now that we were quickly gaining confidence of the local crew - do my best to see if I could enlist their aid in capturing another specimen and preserving it for science. Whether or not it had anything to do with my momentous discovery with the gambusia, I didn't know, but I suspected that it had.

During the gaps in filming throughout the day I had become more and more friendly with the crew. Locals to a man, they seemed childishly flattered in my inept attempt to learn their language, and despite the fact that they spoke little English and I spoke very little Spanish we were joking and laughing like old friends. Some of them actually came from Canóvanas, and I was hoping that I would be able to enlist their aid in the next stage of my investigation.

We drove down towards the town, about half an hour before darkness. I walkie-talkied the others, and asked if they would pull into a lay-by just ahead and wait for us. Nick and I had to make a short detour and I promised that I would explain everything back at the hotel. I found the turning that I was looking for without much difficulty, and got Nick to take the jeep down the narrow road to the bridge I remembered to so well. This was not only the place where my previous director had made a big ass of himself falling in the river, and was not just the location where Graham and I had exhibited great solidarity with each other by refusing, even when separated and goaded into it, to criticise each other for the benefit of the cameras.

It was the place where Graham was being interviewed, and I had gone fishing and had caught a lamprey. Apart from gambusia, the other fish species that I know most about is the brook lamprey (*Lampetra planeri*).

We drove down the hill and parked at a small roadside restaurant - coincidentally - situated by the side of a river where, in 1998, I made a discovery which could - I believe - have changed the face of vertebrate biology as we know it.

When my family left Hong Kong in the spring of 1971, we moved to the idyllic village of Woolfardisworthy in North Devon. For the next few years, I lived the childhood of a character from an Enid Blyton story, and roamed the local countryside in search of the local wildlife.

My father had become close friends with John Raffe - a farmer who lived at Town Farm in the village. He was kind enough to tell my brother and I that we could have the run of his farm, to come and go as we wished, and to explore wherever we wanted. Just outside the village, there is a hamlet called Venn, and at the bottom of the hill there was a small stream, which flowed through John Raffe's land. Despite the fact that now - pushing fifty, and living once again in Woolfardisworthy - I see it as an unprepossessing drainage stream never more than two foot across, and a couple of feet deep, it is still - in my mind's eye at least - Venn River; the place where I made dams, played pirates, sailed my homemade boats, and caught fishes.

One day during the summer of 1972, my best friend David Braund and I were fishing down there. I had brought back from Hong Kong a large fish tank, which had been placed in the garden outside the wall of my bedroom. In it, David and I kept a wide range of the local aquatic fauna, and on this particular day we were hoping to catch some bullheads *(Cottus gobio)* so we could watch them breeding.

Bullheads are strange little fish, which are highly territorial. During the breeding season, the males exhibit striking colouration and develop large knobbly lumps on their heads. David and I found these peculiar little fish to be of immeasurable interest, and although we never succeeded rearing the young satisfactorily, had managed to induce spawning in the aquarium on several occasions. It had been somewhat of a disappointment to me that until then, we had only found two species of fish in the Venn River: bullheads, and strange elongate fishes called stone loach *(Noemacheilus barbatulus)*, but on this day - one of the red letter days from my childhood memories - we found a third species.

It was David that caught it; a strange, eyeless, eel-like fish about eight inches long. It was olive in colour, with a row of tiny holes just behind its head. These were its gills, and together with the strange fleshy folds of its mouth, told us what it was. It was a brook lamprey *(Lampetra planeri)*. Because it had no visible eyes, we could see that it was an ammocoete, or juvenile specimen, and we brought it back to the fish tank outside my bedroom window with pride.

Even at the relatively young age of twelve, I had a burgeoning menagerie of creatures. Indeed, soon after I moved back to the village to look after my dying father in the summer of 2005, I was asked to give a lecture about my life and work to the local village society. I opened my address with the lines:

"Some of you may remember me from over thirty years ago. I was a grubby, perennially untidy small boy who wandered around the village catching creepy crawlies, getting covered in mud, and keeping a

collection of weird wriggly things in a shed at the top of the garden. Well, nothing much has changed....."

This brought the appreciative laughter that I had hoped for, but I was telling nothing but the truth. The only difference is that now I can eke out a living from my perennial interest in weird animals. Of all the animals I have ever kept, then and now, I think that the brook lamprey is amongst the strangest. They are fascinating little creatures, and I dearly hope that the CFZ menagerie, which is a little more sophisticated, although equally as ambitious as the one which I had as a child, will one day again include some of these bizarre animals. Over the next few years I made a study of them, both in captivity, and the wild, and although it has been nearly twenty years since I last saw one alive, they are still one of my favourite creatures. There are three species of lamprey found in British rivers - the two largest, the sea lamprey *(Petromyzon marinus)* and the river lamprey, or lampern, *(Lampetra fluviatilis)* both spend large proportions of their life cycle in the sea. I have never been lucky enough to see either of these species in Britain, although I have seen sea lampreys (albeit a land-locked population) in Lake Superior in Canada. Both the sea-going lampreys feed by attaching themselves to larger fish with their circular jaws, rasping a hole in the skin of its prey with its bony tongue. They secrete a fluid, which dissolves the tissues and keeps the blood from coagulating, so that the lamprey can feast upon the vital juices of its prey.

Both sea-going species migrate inland from the sea in order to spawn. Between February and June, both species indulge in complex courtship behaviour. The female attaches herself to a stone on the bed of the river or stream with her sucker-life mouth, and shapes a small pit out by lashing her tail. While doing so she wafts pheromones into the running water and can attract dozens of eager males, all determined to fertilise her eggs (between four thousand and forty thousand). The males fight each other for the privilege of being able to procreate. They attach themselves to her with their suckers and the resulting orgy of violent sex is so exhausting that the adults will die after spawning.

The babies hatch into blind, toothless ammocoetes which live buried in river mud and living off aquatic invertebrates for three to five years, until they metamorphose into the adult form and swim out to sea.

The brook lamprey, however, is a much more degenerate, and primitive, creature, which is practically unique amongst British fish in having an annual life cycle. The ammocoetes live for three to five months in the river mud, but during metamorphosis when they develop eyes and teeth, their alimentary tract degenerates, and thus the desperate race against time for the breeding cycle to be completed is also a race against time before the adults starve to death.

Lampreys have always been obscure little creatures. They are apparently good eating, and I have read that large numbers are harvested for food in Gloucestershire each year. They only ever took prominence on the world's stage once: when Henry I died on 1st December 1135 of food poisoning from eating "a surfeit of lampreys" (of which he was excessively fond) at Saint-Denis-en-Lyons (now Lyons-la-Forêt) in Normandy.

I have been fascinated by lampreys for years, so imagine my surprise when, in January 1998 I encountered one far from my native North Devon in the little river which flows through Canóvanas.

As you might have gathered by now, my feelings about my first Puerto Rico expedition are somewhat mixed. I was younger, naïve, and drunk too much. Graham had great difficulty in dealing with the tropical heat, and the film crew seemed intent on portraying us as loveable buffoons rather than as serious scientists. One of the tactics they employed was the separate us and interview us in depth hoping that one of us would say something derogatory about the other. We were soon wise to this, and never obliged them, but on one afternoon in mid-January we found ourselves sitting on a low concrete bridge which spanned the river. Graham was being interviewed first and, being somewhat bored with the proceedings, I went wandering up the river with a carrier bag of specimen jars and a small dip net to see what I could find. It was surprising how familiar most of the denizens of this little waterway were. There were may fly larvae, caddis fly larvae and stone fly larvae which would not have been out of place in my native North Devon. There were tiny jewelled freshwater crabs that looked identical to all intents and purposes to those I had caught as a child in Hong Kong. I was entranced to see species in the wild that I had only previously seen in tanks at a tropical fish shop, and I spent a happy half hour watching a small shoal of corydoras catfish chuntering across the river bed like miniature bulldozers. Every stone that I turned over revealed a myriad of exciting life forms, but it was as I was approaching one of the deeper parts of the river (maybe two feet in depth) that I made my momentous discovery. I turned over a rotten log which lay submerged at the bottom of the riverbed and up swam a long, sinuous shape. Swooping enthusiastically with my dip net, I caught the creature and was entranced to find something which looked for all the world like a North Devon brook lamprey in the process of metamorphosing from an ammocoete to an adult.

I don't know who was more surprised – me or the lamprey. I held it in my cupped hands for several minutes and admired its sucker-like mouth, its compact thinness, and its row of gill holes. Then a call from the director summoned me back to reality, so I released the lamprey, went back to be filmed, and forgot about the whole incident.

It was only when I returned to the UK that I realised what a magnificent opportunity I had missed. According to accepted scientific wisdom, not only are there no lampreys in the Caribbean, but there ain't no such thing as a tropical lamprey anywhere. Lampreys are found exclusively in temperate regions (albeit both in the northern and southern hemispheres) and for reasons that no-one seems to understand, they have never evolved to live in warm waters. My lamprey would seem to have been a first.

I have already admitted that far from visiting Puerto Rico purely in order to make a documentary about the chupacabra, I had - to coin one of Nick's phrases - a covert agenda; to catch some of the peculiar flying saucer-shaped snails which I have described in an earlier chapter. However, I had another mission in mind when I agreed to return to the Island of Paradise. I wanted to catch another lamprey.

Nick knew this of course. He had heard my lecture, and had read my books, and I think he shared my grief when I found that the beautiful slow running river full of fish, aquatic insects, tiny red crabs scuttling about like little jewels beneath the crystal water, and weird dark blue squat crayfish like misshapen lobsters with a bad attitude which lurked beneath the round flat stones, had been destroyed. The whole thing had been concreted over and the once beautiful watercourse now flowed through concrete culverts as part of a drainage system for over-flooded fields. We drove a little bit upstream to where the river was a river again. The crystal waters were now the colour of stale Ovaltine; it had rained badly over the previous week, and so this could have been just the silty run off from the rainforest mountains above us, but it looked suspiciously like pollution to me.

In a stark contrast to my earlier feelings of triumph, I felt like a little part of me had died, and crestfallen we drove back to rejoin our comrades.

In a somewhat sombre mood we drove back towards Canóvanas. This time we knew the way good enough not to drive in convoy, and so Nick and I made our own way back. As had become our custom, we played raucous punk music as we drove, and there was something highly surreal about listening to *Sham 69* singing about a borstal breakout as we drove through the plantain groves and peasant farms of the Puerto Rican lowlands. As I pointed out to Nick that particular band had been bad enough back in 1976, and were just an anachronism now. I doubted whether anyone much in England was listening to them at that particular moment, and we agreed that we were probably the only people on the entire island of Puerto Rico with that particular selection of music playing.

As we entered San Juan we must have taken the wrong turning, as we got lost. We drove round and round the weird streets in the residential sector. Outside the tree frogs were shrieking and the big gold harvest moon shone implacably down upon us. It turned out that everybody we asked for directions, either could not speak English or was a visitor, but eventually we got instructions from a shabby looking American hippy and tired after our day's exertions we got back to the hotel just before 9.00pm

Going straight into the bar, we ordered dinner - Spanish omelettes and a selection of local salads. Nick was especially tired after having driven all day and so he went to bed early. The rest of the crew were nowhere to be seen, and I was just beginning to think about following Nick's example, and going back to my room to see what was on television and have a deep meaningful conversation with my snails, when a polite looking man in his mid-thirties - with dark skin and an accent which I couldn't place - came up. *"I couldn't help overhearing your conversation"*, he said. *"You are here looking for monsters, I believe. Do you know anything about the yeti?"*

I told him that I did, and we started to talk. It turned out he was Nepalese and although had never seen the fabled man-beast of the Himalayas for himself he knew people who had and his family had been steeped in the lore of such things, but this wasn't the biggest coincidence of the evening. It turned out that he was a British soldier – a Ghurkha, and he was stationed at Catterick Camp in Yorkshire. On hearing this, aghast, I must have looked visibly surprised because he asked me why.

I told him that my brother was one of the regimental chaplains at Catterick Camp, and his jaw dropped. *"You don't mean the Reverend Downes do you?"* he asked. I nodded. It turned out that not only was my new friend a Ghurkha but he was one of the very few Christian Ghurkhas in the British Army and that my brother, in addition to his other duties, was Chaplain to this very regiment.

This called for a drink, and as we drank and chatted we were joined by another couple that were staying at the same hotel. They were from Seattle and it turned out that they had witnessed what they believe was bigfoot. By this time I was used to strange things happening, and so spent a happy evening chatting and drinking and joking with my new friends until the beautiful barmaid with skin like milky coffee told us that she was shutting up the bar, and that it was time for all of us to go to bed. Even chucking everyone out, she did so sweetly and prettily, that I am certain that she made over a hundred bucks in tips in about three minutes!

We said our goodbyes and as I walked the two hundred yards up the road to the apartment block where I was staying, I listened to the crickets and uttered a silent prayer to the Almighty to thank Him for having made such a wonderful and peculiar world.

The Margarita Diary,
Chapter 5 Comments:

There's something special about driving around in an open-top jeep in a place like Puerto Rico with one of your best friends, with the wind in your hair (for those who have hair...), and in hot pursuit of the unknown, while the ear-splitting tunes of Joey, Johnny, Dee Dee and Marky echo around the woods. Approximately seven months earlier, I had the same sensation when Jon, conspiracy author Kenn Thomas, good friend and fellow writer Greg Bishop, and base-invader and crop-circle maker Matthew Williams, and I headed out into the wilds of the Nevada desert when the sterile confines of a certain Las Vegas UFO gig got far too much for us.

This time, however, rather than just a few hours on a Sunday afternoon, we had a whole week ahead of us. I have to say that the interview we conducted at our first port of call was somewhat surreal. I had read enough about the chupacabras and its supposed predations to know that we were in for some high-strangeness. Yet, for all I had seen and done, it was still unsettling and weird to speak with someone whose animals had allegedly been drained of blood via the classic method of two puncture marks to the neck – and in an isolated and run-down little Puerto Rican village, too.

Whatever the truth of the affair, I knew that barely an hour or so into our adventure we were making good headway. And, of course, no expedition of this type would be complete without an excursion into the darkened depths of a shadowy old cave. That a bat decided to piss on my head while we were in there only made things more memorable. I decided not to bother with rabies injections of a type that Ozzy Osbourne was forced to undergo after his own legendary encounter with a bat; and instead hoped that the little pisser wasn't rabid, and that I wouldn't wake up the next day like one of the frenzied souls from *28 Days Later* or the spectacular 2004 remake of *Dawn of the Dead*. Needless to say, I didn't.

Of course, I knew that all of this would serve as good fodder for Jon's planned book on our trip around the island, and so I merely wiped my head with my bandana, swore at the offending beast and his or her brethren and continued roaming and filming. And a crew of a dozen, led by the good Mr. Downes himself, laughed heartily! **NR**

Chapter Six
Saucer smears

Sometime during September 1957 "something" crashed into the hillside deep inside the El Yunque rainforest. On no less than five separate occasions during my two visits to the Island of Paradise local people told me of the incident and described what had crashed as a UFO. They were, of course, in the truest sense of the word, perfectly correct. If indeed something did crash onto the forest-covered mountain that night nearly half a century ago it was:

a. an object
b. flying
c. unidentified

However, there is little if no evidence to say that it was a flying saucer. It is one of the mildly irritating things about contemporary forteana that the term UFO is used in a cavalier fashion, and instead of merely meaning an unidentified flying object, it has become widely regarded as a synonym for "alien spacecraft". This is most unfortunate, because this *idée fixe* - which has no hard evidence to support it - dominates the entire study of the subject, and threatens time and time again to overwhelm it.

It is generally believed that the universe is infinitely large. If this is so, then it is a statistical certainty that other intelligent beings have evolved *somewhere* else in the cosmos. However, it is also almost certain - unless they have developed some way of meeting the limitations imposed by the speed of light – that, except by accident, it is practically impossible for them to reach earth.

During the first week of July 1947, William "Mack" Brazel, a rancher from New Mexico, discovered a large amount of unusual debris scattered widely over his ranch near Corona, New Mexico, about 75 miles northwest of Roswell. Debris - and according to some reports, tiny bodies - were taken away by the Military.

An industry was born.

Nearly sixty years later, Nick Redfern published a book called *Body Snatchers in the Desert*. In it he presents a compelling argument that the so-called UFO crashes at Roswell and other locations in the American desert states in the late 1940s were the results of covert experimentation carried out by certain factions within the American Government. [1]

It is a matter of historical record that when the Nazi regime collapsed in the late spring of 1945, that there was an ungodly rush by the American, British, French and Russian Governments to grab a hold of the secrets of the Nazi technological advances.

Operation Paperclip was the codename under which the US intelligence and military services extracted scientists from Germany, during and after the final stages of World War II. The project was originally called Operation Overcast, and is sometimes also known as Project Paperclip. Of particular interest were scientists specialising in aerodynamics and rocketry (such as those involved in the V1 and V2 projects, the abortive V3 supergun project, and other secret weapons such as the various *Wunderwaffen* or super-weapons, most of which did not get past the drawing board stage), chemical weapons, chemical reaction technology and medicine. These scientists and their families were secretly brought to the United States, without State Department review and approval; their service for Hitler's Third Reich also disqualified them from officially obtaining visas. An aim of the operation was capturing equipment before the Soviets came in. The US Army destroyed some of the German equipment to prevent it from being captured by the advancing Soviet Army.

The majority of the scientists, numbering almost 500, were deployed at White Sands Proving Ground, New Mexico, Fort Bliss, Texas and Huntsville, Alabama to work on guided missile and ballistic missile technology. This in turn led to the foundation of NASA and the US ICBM program. Much of the information surrounding Operation Paperclip is still classified, and many scientists, most notably Werner von Braun, the German rocketry genius, who were at least as guilty of war crimes as many of those prosecuted and hanged at Nuremberg escaped the hangman's noose and were spirited away to America where they worked on the nascent space programme.

Many of the alleged *Wunderwaffen* exist in that twilight world between science, history and folklore, and include a number of projects aimed at producing sub-orbital spacecraft, and ICBMs which would - had they actually been completed - have certainly turned the tide of history a different way. However, it is tempting to speculate that the rapid advances in both US and Soviet technology in the immediate post-war years were down to timely misappropriation of German technology.

The American Government was justly proud of having landed men on the moon in July 1969. However, this undoubtedly impressive achievement would not have happened if it had not been for some gruesome and totally illegal experiments carried out in Nazi death camps by Dr. Mengele and his ilk. They carried out experiments to test the tolerance of the human body to high altitude, low temperature, low pressure and rapid decompression.

Half a world away, the Japanese had also been carrying out secret weapon research during WW2, although - as anyone who has borne witness to the amazing technological advances in Japan over the last 50 years will not be surprised to learn - many of these researches were undertaken into remote controlled bombs, surface-to-air missiles, and smaller items of technology. There was a Japanese nuclear weapons

---

1. Most of the information on Roswell, the Japanese military experiments and Operation Paperclip in this chapter comes from *Body Snatchers in the Desert: The Horrible Truth at the Heart of the Roswell Story* by Nicholas Redfern ISBN-13: 978-0743497534, but I also consulted *The Day After Roswell: A Former Pentagon Official Reveals the U.S. Government's Shocking UFO Cover-up* by Philip J. Corso and William J. Birnes ISBN-13: 978-0671017569 and *UFO crash at Roswell* by Kevin Randle and Donald Schmidt **ISBN-13:** 978-0380761968

programme, but despite claims that they actually completed one bomb and detonated it a week before the bombing of Hiroshima, they were never brought into production. However, in the field of biological research, they were second to none, and their research - both in scope, and brutality - probably surpassed that of the Germans.

The notorious Unit 731, presumably with the full knowledge of the Japanese Government was carrying out similar experiments on prisoners of war and mentally and physically handicapped children and adults. Unit 731 was a secret military medical unit of the Imperial Japanese Army that researched biological warfare and other topics through human experimentation during the Sino-Japanese War (1937-1945) and World War II era. Disguised as a water purification unit, it was based in Pingfan, near the city of Harbin in northeastern China, the region which was then part of the puppet state of Manchukuo. It is estimated that over 3,000 Chinese, Korean, and Allied POWs were killed in the Unit 731 facilities. Many more people died in field experiments directed by Unit 731, but there is no well-established number.

A special project code-named *Maruta* used human beings for experiments. Test subjects were gathered from the surrounding population and were sometimes known as "logs".

This term was the result of the feeling of the scientists that killing a prisoner was the same as cutting down a tree. The test subjects ranged from infants, to old people, to pregnant women along with the baby. Many experiments were performed without the use of anaesthetics because it was believed that it might affect the results.

Live vivisections were performed on prisoners infected with various diseases; scientists would remove organs to study the effects of the disease on the human body. Prisoners were amputated limb by limb to study blood loss. Arms were cut off, and reattached to opposite sides. Limbs were frozen and sawed off. Stomachs were surgically removed and the oesophagus was reattached to the intestines. Parts of the brain, lungs, liver, etc., were taken out.

There were other units besides Unit 731, which serves as a general term in describing the Japanese biological warfare program. Other units include Unit 543 (Hailar), Unit 773 (Songo unit), Unit 100 (Changchun), Unit 1644 (Nanjing), Unit 1855 (Beijing), Unit 8604 (Guangzhou), and Unit 9420 (Singapore). The acts of Unit 731 are one of many major war crimes committed by the Imperial Japanese Army from the occupation of Manchuria in 1931 to the end of World War II in 1945.

After these laboratories were destroyed by the Japanese to hide their activities, many of the scientists involved went on to prominent careers in politics, academia and business. In a similar move to that in occupied Europe, The United States granted amnesty, allowing these scientists to go unprosecuted in exchange for their experimentation data.

Redfern's hypothesis is that when American troops liberated a secret Japanese research station carrying out experiments with high altitude passenger-carrying balloons, as well as bringing the scientists, the hardware and documentation back to the United States, they - in one of the most shameful incidents of modern American history - brought back several human experimental subjects: young men and women with progeria, in order to continue these experiments.

Progeria [1] is an extremely rare genetic condition which causes physical changes that resemble greatly

---

1. In my younger days I was a nurse for the Mentally Handicapped. Indeed I have the right to include the letters RNMS after my name. I don't because it is a professional qualification that I don't use, and that I am not particularly proud of. However, during my student days I made a study of progeria - something which came in very useful whilst writing this chapter which was based largely on my student notes.

accelerated aging in sufferers. Most children with the disease die around 13 years of age. Symptoms generally begin appearing around 18-24 months of age.

The condition is distinguished by limited growth, alopecia and a characteristic appearance with small face and jaw and pinched nose. Later the condition causes wrinkled skin, atherosclerosis and cardiovascular problems. Mental development is also usually affected.

Individuals with the condition rarely live more than 16 years; the longest recorded life-span is 26 years. The development of symptoms is comparable to aging at a rate six to eight times faster than normal, although certain age-related conditions do not occur.

I agree broadly with Nick Redfern's hypothesis, and believe that eyewitness reports of having seen little bodies on the ground near a crashed flying saucer in Roswell, New Mexico in July 1947, are substantially correct. However, what they were seeing was not the aftermath of a crashed alien spacecraft. What they were actually seeing was a crashed high altitude balloon built to the specifications designed by the Japanese for use in action in 1940, and what was left of its deformed human cargo.

It could well be argued that the Roswell incident is the cornerstone of most of modern UFOlogy, and once that has been taken out of the loop (which if you believe Redfern's theory, it has), then much of contemporary UFOlogy begins to look decidedly dodgy.

Majestic-12 (sometimes written simply as MJ-12 or MJ-XII) is the codename of a secret committee presumed to have been formed in 1947 at the direction of U.S. President Harry S. Truman, in order to investigate UFO activity. This alleged committee is an important part of the UFO conspiracy theory.

The only evidence of MJ-12's existence is a series of purportedly genuine and secret documents, which have been the subject of much debate.

Published copies of the MJ-12 documents which state that *"the Majestic 12 (Majic 12) group ... was established by secret executive order of President Truman on 24 September, 1947, upon recomendation by Dr. Vannevar Bush and Secretary James Forrestal."*

The existence of MJ-12 has been denied by the United States government, which insists that documents suggesting its existence are hoaxed. The FBI investigated the documents, and concluded they were forgeries. Opinions among UFO researchers and enthusiasts are divided: Some argue the documents may be genuine, others contend they are phoney, due primarily to errors in formatting and chronology.

If, however, the events of July 1947 *did* take place, then the only way that MJ-12 could possibly have existed was as a source for disinformation.

If, indeed, as Redfern claims, certain rogue elements within the United States Military-Industrial complex were actively involved in what is arguably the most shameful episode of modern American history, then it becomes more and more logical that the MJ-12 documents and even the so-called original alien autopsy video, (it is now widely known that the film that surfaced in the mid 1990s in the UK was actually - as many of us have said for years - faked by Ray Santilli and his chum) were all parts of a complex campaign of disinformation to cover up the truth of what really happened.

After all, it would not have been the first time that the United States security services had used superstition, and claims of what are usually lumped together as "the paranormal", as a tool of psychological warfare. In the Philippine-American War of 1899-1913, American soldiers on several occasions mutilated the bodies of captured Philippino natives to suggest that they had been victims of an *aswang* - the local example of the global vampire mythos [1]. It has also been suggested that similar events took place in the sixties and seventies during covert operations in Central America. If we take on board the hypothesis that such actions have been regular weapons in the US disinformation arsenal, then the UFO phenomenon worldwide takes on an entirely different appearance.

Puerto Rico has been a hotbed of UFO activity for many years, and the belief in OVNIs [Objeto Voador Não Identificado] is ingrained deeply into the consciousness of those who live there. In my travels around the world I do not think that I have ever found a place that sports quite so much UFO imagery. There are UFO bars, UFO t-shirts, UFO toys and UFO songs.

This widespread cultural belief has sparked off a minor industry elsewhere in the globe, with many researchers (who really should know better) writing books packed to a greater or lesser degree with nonsense, claiming that the island is home to a number of secret UFO bases from which alien greys sally forth to do unpleasant business with the local representatives of the American Government. At the time of my first visit to the island in 1998 the world - and Britain in particular - had just been through an outpouring of media interest of the subject of things unidentified and flying, and many people - myself included - had somewhat cynically, made quite a lot of money out of it. I was, therefore, feeling somewhat jaded about all matters saucer-like, and so when I first heard about the 1957 incident I was tempted to consign it to the dustbin of history together with other recent Puerto Rican incidents, like the affair of the so-called alien foetus I described in an earlier chapter - which I had investigated for myself, and found to be more or less fallacious.

Almost as soon as we arrived on the island in 1998 we were told stories of UFO activity. Rosario, our interpreter and Ms. Fixit arranged for us to meet a family in the Rodriguez sector of Dorado township. These people - apparently - had seen a full-blown chupacabra emerge from a flying saucer. Stifling a yawn, we wearily agreed to go and visit them. So, on our first full day's investigation on Puerto Rico in

---

1. My reference for this rather nasty story is Nick Redfern who cited two sources: Edward Lansdale in *The U.S Air Force Biography*, and an article called *Psywar Terror Tactics* by Jon Elliston at www.parascope.com and http://karws.gso.uri.edu/Marsh/Bay_of_Pigs/psy.htm It reads:

Many early U.S. psywar operations were conceived by a famous clandestine commander, Air Force Brigadier General Edward G. Lansdale (1909-1987). A firm believer in the efficacy of "psychological operations" (or PSYOP, for short - the military's term for propaganda), Lansdale was a pioneering psywarrior. Lansdale believed that the key asset of the psychological combatant is a thorough understanding of the target audience's beliefs and values. The mores and myths that shape a society's culture, he argued, must be exploited if a psywar campaign is to be effective. Lansdale applied his strategy ruthlessly in the Philippines, where he served as the CIA's chief operative during the early 1950s counterinsurgency campaign against the country's Huk rebels.

"To the superstitious, the Huk battleground was a haunted place filled with ghosts and eerie creatures," Lansdale later wrote. One of his favorite psywar stunts "played upon the popular dread of asuang, or vampire" to drive the guerrillas from Huk-held territory:

"A combat psywar squad was brought in. It planted stories among town residents of an asuang living on the hill where the Huks were based. Two nights later, after giving the stories time to make their way up to the hill camp, the psywar squad set up an ambush along the trail used by the Huks. When a Huk patrol came along the trail, the ambushers silently snatched the last man of the patrol, their move unseen in the dark night. They punctured his neck with two holes, vampire-fashion, held the body up by the heels, drained it of blood, and put the corpse back on the trail. When the Huks returned to look for the missing man and found their bloodless comrade, every member of the patrol believed that the asuang had got him and that one of them would be next if they remained on that hill. When daylight came, the whole Huk squadron moved out of the vicinity."

January 1998 we found ourselves driving off in search of them. Luckily, as it turned out, the instructions that we had been given were entirely wrong, because although we never found the Rodriguez family we did find something else which was to change the entire course of my research over the next ten years.

I do not know how she had managed it, but Rosario had arranged for us to be given a police escort during our first full day outside San Juan. It was, as Graham said, his first time in the *front* of a police car, and - to my great pleasure - we found that our mentor of the day, a charming police officer called Reuben was one of the few people we would meet in authority who was to take our quest seriously.

We soon found out why.

Although from a Puerto Rican family, Reuben had been brought up in New York, and his English was excellent. Unlike Graham I had ridden in the front of a police car before whilst working as a community nurse in the early 1980s, so I was not fazed. What *did* alarm me was to be so close to such an impressive battery of firearms as those carried by Reuben and his pal. I was reassured to see that despite the handgun on his belt, the tear gas pistol (and what looked like a sub-machine gun) in the car, and the suspicious bulge which I presumed was a shoulder holster, Reuben had a kind look on his weather-beaten olive brown face, and a mischievous twinkle in his dark eyes. After driving around the Rodriguez sector fruitlessly for an hour and a half, we stopped in a lay-by for a cigarette, and Reuben asked us about the background to our quest. I explained that although I was a firm believer in the chupacabra, and that we were certain not coming all the way to Puerto Rico in order to denigrate local belief systems, I was, and am, a scientist and that I wanted to find a scientifically acceptable explanation for the wave of grotesque vampiric activity which had spread across the island in recent years.

He nodded his head in agreement, took a deep drag on his cigarette and said, *"Yeah, I agree totally. This stuff is real. I have been to enough of these attacks in the last few years to take them seriously, but there is an awful lot of horseshit talked about them."* His colleague, sitting in the back seat just grunted. He was obviously a man of few words.

I very nearly made a fatal mistake at this juncture. I assumed that because Reuben and I were getting on so well, that we would have similar mindsets, and therefore a similar interpretation of the problem. I said something in passing about the Rodriguez family and their claim to have seen a chupacabra alighting from a flying saucer. I was just about to make a stupidly deprecating remark, when - without realising it - Reuben saved me from a potentially disastrous social situation.

"I don't know about that," he drawled, still taking deep lungfuls of cigarette smoke, *"But OVNIs are real. My daddy even saw one crash!"*

This was obviously not the time or the place for me to parrot as one of my explanations about UFOs being a psychosocial phenomenon rather than something physical, and so I said something non-committal, and Reuben's mate grunted encouragingly for me to continue.

It turned out that Reuben's father before him had been a policeman in Puerto Rico, and had moved to New York in 1958 to get married and start a family. Apparently, about six months before he left for the Big Apple he had been in his patrol car driving along the narrow road that leads through the high grassland plateau above Canóvanas and winds up towards the foothills of El Yunque herself. Late one September night, a great flash split the ebony sky, and Reuben's father saw a giant fireball hurtle out of the sky and into the forested mountainside above him. He called for assistance, and drove as close as he could to the impact site in order to investigate.

He was still negotiating the narrow mountain roads that lead up the steep mountainside, when he became aware that he was not alone. Above and around him were lights, and he could see some of the forested hillside illuminated as brightly as if it were day. Four helicopters had appeared on the scene and were combing the area with powerful searchlights. He drove as far as he could, parked his patrol car, and prepared to continue his investigation on foot. It was too dark and regretfully he was forced to turn back. The next day he returned there to find the area swarming with soldiers. These were not the local militia or National Guard. These were tough combat troops from the American mainland, who told him in no uncertain terms that the area that he wanted to investigate was now classified, and that as a civilian he was not allowed to proceed any further.

"And you know what?" he said. "That part of the hillside was taken over by the Army for the next twenty-five years."

Reuben had heard this story as a little boy, and although as an adult he had returned to the island of his forebears, as a policeman, he had never forgotten the story. He told us - roughly - where the crash site was, and a few days later, together with a film crew from Channel 4, we went to see if we could find it.

As anyone who has read my first book about Puerto Rico will know, by this time in the expedition, relationships between the director and me were beginning to fray. Essentially we had completely different agendas. Norman Hull is an undoubtedly fine director, a brief foray on to the Internet will tell you how at least one of his films was entered at Cannes, and how he is generally thought of as one of Britain's leading documentary makers. However, his agenda was to make an entertaining film, whereas mine was not only to hunt for the chupacabra, but to follow whatever leads the patron saint of cryptoinvestigative methodology would throw in my path. Although I tried to remain focused during my investigations, sometimes something so far-out comes along, that it would be stupid not to follow it up. One such instance during our first trip to the island of paradise had been my impromptu autopsy on the dead chicken. It hadn't been in the script, therefore Norman was unwilling to waste time on it. That caused an enormous stand up argument between us, and a few days later we were to have another one when we appeared on a syndicated early morning radio show. It was a phone-in and I was being asked questions about the Florida Skunk Ape. These I was happy to answer, but Norman felt that I was becoming unfocused, and that I should have refused to talk about anything apart from the direct aims and objectives of our quest. The relationship between him and me deteriorated to such an extent that - at one point - I was even tempted to walk out and fly back to England. I was certain that any attempts on my behalf to take time off investigating a putative UFO crash, would be met with complete hostility. I managed to manipulate the situation, however, whereby we would visit the alleged crash site during our sojourn in El Yunque rainforest, and I told Norman and the rest of the crew that Reuben had recommended this particular location for filming.

Our day in the rainforest did not get off to a particularly auspicious start. It turned out that one needed a permit in order to make a commercial film within some parts of the national park, and so as we had not obtained a permit, and as Norman and his company were unwilling to do so, we went up the mountain soon after dawn. We parked at the crash barrier, and - shouldering our equipment - we trudged deep into the forest. It was eerily quiet. In my earlier book I commented on the fact that in this particular location we found practically no animals. I have visited rainforests in Australia, Africa, and South-east Asia, and invariably every rotten log is a Lilliputian universe populated by woodlice, ants, centipedes and millipedes, beetles, and even small reptiles and amphibians. Here it was very different. The forest seemed dead. And at the risk of misquoting Sigmund Freud, there a log was just a log, and an extremely uninteresting one at that.

We trudged into the forest for over an hour, along little winding paths mottled with little pools of sunlight, and the dark shimmering green of the undergrowth. To my great annoyance Norman wanted to make it appear that I was suffering more than I actually was. Despite being seriously overweight, and suffering from an impressive array of physical and mental health issues, I had been coping quite well with the tropical heat. After all, the climatic conditions in Puerto Rico were not much different to those of the Hong Kong of my childhood, and for the first time in three decades or so I felt quite comfortably at home. However, Norman, using the same degree of empathy with my weight problem which he showed later in the trip when he insisted on filming me squeezing myself in and out of smaller and smaller cars, insisted that I walked up and down, up and down, and up and down again until I was panting for breath, and salty sweat ran down my face and into my eyes. Only then, once I had artificially exhausted myself for the sake of Norman's artistic integrity, was I allowed to rest.

Eventually we came to a big clearing where the path became narrow and on one side disappeared altogether into a huge saucer-shaped arena. This, according to Reuben at least, was where the UFO had crashed. Admittedly, there was a huge indent in the side of the mountain. No trees grew there, and it did look for all the world as if some huge object had crashed into the mountain, scooping out trees and vegetation and leaving a bare area intermittently covered with patchy grass. I have seen such places before on the clay-covered sides of tropical mountains, but in these cases, the reason for them is obvious. Quite simply the clay soil is not strong enough to support the weight of huge forest trees, and so once they reach a certain size they slip down the side in impressive landslides following the first major rainstorm. However, there was no sign of this happening. Also, a rainforest is a hotbed of animal and plant life. An area that could not hold sizeable trees would nevertheless be covered with ferns and small undergrowth, but there was nothing of the sort here. It was just a huge brown slippery saucer-shaped indenture out of which a tiny mountain spring listlessly trickled.

With our binoculars we could follow the stream down the hillside until it joined one of the larger rivers which eventually formed the headwaters of the mighty Canóvanas River.

Just over six years later I returned, and although I had not forgotten Reuben's tale of UFO crash in the midst of the mighty jungle, it had been put securely into my mental filing cabinet for retrieval at a later date.

In November 2003 I had been invited to lecture at a UFO Crash Retrievals Convention in Las Vegas. I agreed to attend the conference with my tongue firmly in my cheek. The only reason I decided to go was because I had not seen Nick Redfern for a couple of years, and he - and other friends of mine, such as Peter Robbins, co-author of *Left at Eastgate* - were going to be there, and I fancied leaving the cold grey English winter behind me and spending 10 days as the guest of a family of exceedingly rich Americans in Sin City.

However, there was a problem. As I have stated over and over again, I do not believe in the extraterrestrial hypothesis, I do not believe in flying saucers, as far I am concerned the MJ12 documents are palpable fakes, and to be quite honest I don't give a damn what - if anything - crashed into the Corona desert that day in July 1947.

But the conference was purely on the subject of crash retrievals, and was organised and attended by the most hardcore of true believers. What was I to do?

After much discussion with Nick, who by this time had become so disillusioned with the whole thing that he - too - had come to see this conference merely as an excuse to get two of his best friends over for a junket in Vegas. I gave a talk loosely based around my experiences with Reuben during my 1998 trip to the island. I was billed as follows:

Island of Paradise

"Jonathan Downes — Puerto Rico UFO Crashes

Jonathan Downes will be speaking about a 2-month expedition he made to Puerto Rico with a British TV crew in search of both UFOs and the Vampire-like Chupacabra that is rumored to haunt the forests of Puerto Rico. Downes will be discussing information he uncovered and cases he investigated relating to reports of UFO crash incidents in the Puerto Rican jungle and accounts of Chupacabra bodies recovered under cover of the utmost secrecy by the U.S. military. "

What the punters saw, was most definitely - however - not what they got. The write-up in the programme made me seem like some shadowy whistleblower intent on uncovering the holes in the US Defence Department's UFO strategy, when in fact I am - as I hope has come over in this narrative so far - no such thing.

I know that the conference organisers were not at all happy with the contents of my presentation. Most of the other speakers on the bill were adherents to the most hardcore and fundamentalist doctrines of UFOlogy. The loudest mouthed and most vociferous of them believed implicitly in the voracity of the MJ12 doctrines. They sat in earnest little huddles in the corner of the bar discussing how Truman and Eisenhower ("good old Ike", as the more staunchly Republican of them were to refer to him), had founded Majestic 12 in order to deal with the alien menace. Although the more liberal speakers there, such as Nick and Pete, Kenn Thomas from *Steamshovel Press*, and Greg Bishop – who we met a few chapters ago; a pleasant young man from California who spent most of the time in the bar with me talking about psychedelic rock music - may not have believed in the orthodoxy of the gospel according to Stanton Freidman, but they were wise enough not to rock the boat.

Then there was me.

I was about as welcome within this coterie of true believers as a pork butcher at a Bar Mitzvah. I gave a talk based around my theories, which are laid out in this book, was politely applauded, and roundly ignored. Even the Las Vegas newspapers - who misquoted me to an obscene extent - were more interested in the size of my stomach, and the length of my hair, than in anything that I truly had to say:

```
"One presenter whose lecture was more than a little off the beaten path was
British cryptozoologist Jonathan Downes. Downes has no scientific training
(does that make him a crypto-cryptozoologist?) and has been a full-time profes-
sional writer since the 1980s, specializing in science fiction and speculative
works on alleged paranormal incidents, like the appearance of an "Owlman" in
Cornwall, England. His presentation dealt with his search for the "chupacabra,"
a monster that supposedly sucks the blood and vital organs from Puerto Rican
farm animals, leaving only a puncture wound.

While entertaining, Downes' presentation wasn't convincing. The imposing-
looking writer, who stands well over 6 feet, is of exceptional girth and has
wild longish hair and a colorful way of speaking, showed slides of photos he
took in Puerto Rico. He claimed these photos showed chickens mutilated by the
chupacabra, warning the audience that the pictures might not be for the faint
of heart. However, what he showed were piles of feathers that might have been
chickens, with dark spots on them that might have been holes.

Downes also showed slides of a lamprey and a disfigured guppy, claiming that he
```

found these fishes in Puerto Rico, a place where they were not native. He hypothesized that these creatures evolved very quickly, in the 40 years since a purported UFO crash in Puerto Rico's rain forest.

"The UFO crashed near a stream, and contaminated the water, changing the ecosystem of the area," Downes asserted.

Downes said he accepted the television research assignment to Puerto Rico as a "lark," not believing in the chupacabra so much as wanting to escape a dreary England winter. After his trip, he said he believed in the monster.

"I'm going to leave you with this," he said, and paused dramatically.

"There are still some very primitive peoples in the rain forest of Puerto Rico," he continued. "What if these people, for 40 years, have been drinking the contaminated water - and reproducing?" With that, he went to an image of a two-legged reptilian figure that looked like the twin of the monster in The Creature from the Black Lagoon. There were gasps from the audience.

We have seen the chupacabra, and the chupacabra is us."

Hmmmmmmm. As anyone who has managed to follow this narrative so far will attest, I don't believe anything of the sort!

What I *actually* said was that there was some evidence that *something* was causing Progeria-like symptoms in a village of people who lived by a lagoon near the mouth of the Canóvanas River. Although I had not visited this village, Reuben had told me of the "place, where the children looked like aliens" and in the intervening years I had heard several other rumours that there was a place in the vicinity where something appeared to be wrong with the genetic makeup of the inhabitants. I told this story, and flashed up the famous Jorge Martin image of a chupacabra, which had become so famous in the mid-1990s.

After all, the idea of something falling from the sky, that can cause illness, if not actually effect the genetic makeup of animals and men, is hardly unique. Interestingly, three years later, as I sit in my study in rural North Devon, taking advantage of the last evenings of the year when it will still be warm enough to sit with the door open, I received an email from Nick, with a news story that he thought would be of interest to me.

Mon Sep 17, 11:23 PM ET

LIMA (AFP) - Villagers in southern Peru were struck by a mysterious illness after a meteorite made a fiery crash to Earth in their area, regional authorities said Monday. Around midday Saturday, villagers were startled by an explosion and a fireball that many were convinced was an airplane crashing near their remote village, located in the high Andes department of Puno in the Desaguadero region, near the border with Bolivia.
Residents complained of headaches and vomiting brought on by a "strange odor," local health department official Jorge Lopez told Peruvian radio RPP. Seven policemen who went to check on the reports also became ill and had to be given oxygen before being hospitalized, Lopez said.

Rescue teams and experts were dispatched to the scene, where the meteorite left a 100-foot-wide (30-meter-wide) and 20-foot-deep (six-meter-deep) crater, said local official Marco Limache.

"Boiling water started coming out of the crater and particles of rock and cinders were found nearby. Residents are very concerned," he said.

A few days later, another news release did the grounds, this time giving a slightly different version of events:

Mysteries remain over Peru meteorite impact
* 00:00 28 September 2007
* NewScientist.com news service
* Jeff Hecht

Conspiracy theorists will be disappointed. The object that exploded and formed a crater that emitted mysterious gases in Peru on 15 September was a meteorite that hit soil where the subsurface water table was high, according to the first official report from geologists who have returned from the scene.

Speculation raged about what caused the crater, found in the Peruvian town of Carancas, near the Bolivian border, with a hydrothermal explosion of gas and even a downed spy satellite offered up as culprits.

"The mysterious gases were steam. It was a rock that fell out of the sky and made a hole in the ground. End of Story," says Lionel Jackson of the Geological Survey of Canada in Vancouver. But some questions still remain.

The meteorite came from the north-northeast and was bright enough as it streaked over the city of Desaguadero - which lies 20 kilometres north of Carancas - that many residents there clearly saw it at 1145 local time. Witnesses did not see the fireball break up in the air, but people up to 20 kilometres from the crater reported hearing an explosion - presumably the impact. Windows were shattered at the local health center a kilometre from the impact site.

The space rock hit a region of soft red soil a few metres thick at an elevation of 3.8 kilometres in an area that had been covered by Lake Titicaca during the ice age. A report by Luisa Macedo and José Macharé of the Peruvian Institute for Geology, Mining and Metallurgy describes a crater measuring 13.3 by 13.8 metres, with a rim a metre above the original soil level.

Ground water quickly filled the crater, and witnesses reported that it was boiling. By the time Macedo arrived 36 hours later, she saw "turbid brown" water reaching to within a metre of the original soil level. The crater looks like it was formed in water-saturated soil, says Peter Schultz of Brown University in Providence, Rhode Island, US.

Meteor fragments found at the site were "fine-grained, light grey, fragile rocky material, with disseminated iron [particles] of one-millimetre diameter", Macedo and Macharé report. Thin sections showed tiny silicate spheres found in many meteorites, confirming the rocks came from space.

"The reports about this 'impact' have been confused, contradictory and muddled," says Don Yeomans of NASA's Jet Propulsion Laboratory in Pasadena, California, US. "[This] report is by far the best I've seen."

The Peruvians also reported that early claims of 200 people sickened by the impact were exaggerated. They found only about 30 complaints of ailments such as headaches and nausea, but could not identify a cause. Jackson suspects the victims were simply stunned by "a big explosion in a very quiet area of the world".

Boiling point

Geologists are still puzzling over some details. The fragments looks like stony meteorites, but that type normally breaks apart in the air, and local witnesses saw only a single object. The size of the object that hit also remains unclear - estimates range from the size of a basketball to a few metres across.

"This is probably a fragment from a breakup of a larger object, which broke up at an approximate altitude of 50 kilometres," Yeomans told *New Scientist*. "One would expect a three-metre (original size perhaps) meteorite to hit the Earth's atmosphere a few times each year and sometimes a fragment makes it to the ground."

The speed and temperature of the object also remain a puzzle. Witnesses reported that the crater steamed for half an hour after impact. "It makes no sense that the water was actually boiling," says Clark Chapman of the Southwest Research Institute in Boulder, Colorado, US.

Although a meteorite's surface layer may burn off in the atmosphere, its interior normally remains at the cold temperatures of outer space. The crash site's high altitude reduces the boiling point of water, but only by a little over 10 [degrees] Celsius.

Jackson thinks the kinetic energy of impact could have generated the heat, and is trying to find seismic records of the crash. Schultz says the heat and bubbling might have come from air trapped and heated on the "front" of the meteorite as it sped through the air.

Hmmmmmm.

Over the next few months, during the first part of the long, cold winter of 2007/8 theories, claims and counterclaims flashed and reverberated across cyberspace. The Peruvian Government did their best to try and put a damper on the story, but the speculation continued.

If you are to ignore the theories that the object crashed into the Peruvian countryside was a UFO; I have already explained my position on the subject of extraterrestrial visitors in some detail, and I do not wish to bore you by repeating myself unnecessarily, and if we are to ignore the equally vociferous proponents of the theory that a spy satellite, or some mysterious manmade object - presumably manufactured with some sinister end in mind - crashed to earth in Peru in September 2007, then there are a number of hypotheses which deserve some degree of scrutiny.

Everyone involved, who isn't a complete lunatic, seems to agree that the object which fell to earth was a meteorite. The fragments which have been recovered correspond exactly with known meteorite fragments, and there is nothing amongst the geological evidence to make one suspect anything else. Therefore, the mysterious illness - if indeed, in actually existed - can be attributed to one of four causes.

1. The substance which caused the illness was carried by the meteorite.
2. The substance which caused the illness was buried in the ground, and was unearthed by the impact of the meteorite.
3. The illness was caused by mass hysteria.
4. The whole thing was a coincidence.

Well, as I have already told you, an old and dear friend once told me, there ain't no such thing as a coincidence, and whereas I agree with the probably fictional Lazarus Long that one should "never underestimate the power of human stupidity", the excuse of 'mass hysteria' has been used by so many people, on so many occasions, to explain quasi-fortean phenomena, that it goes against the grain to admit it as a possibility in this case.

Despite claims that the ALH 84001 meteorite which was found in Antarctica in 1984 contains fossil microbes from Mars, and more recently claims in 2005 - which were soon withdrawn -by two NASA scientists, Carol Stoker and Larry Lemke of NASA's Ames Research Center in Silicon Valley, that life may exist today on Mars, hidden away in caves and sustained by pockets of water, there is very little evidence to support any suggestions that the illness in Peru was caused by extraterrestrial biological material. It could, however, have been caused by an - as yet unidentified - mineral, however, and this is one of the more sensible theories that have been put forward. However, possibly the most likely theory, is that the impact of the meteorite broke the surface of the ground, and unearthed some noxious substance, either mineralogical, or biological. The fact that, within a fairly recent timeframe - at least in geological terms, the area had been the bottom of one of the world's largest lakes, would seem to support this hypothesis, and furthermore the hypothesis that the noxious material that was unearthed was probably biological in nature.

Sadly, the whole weekend at the Crash Retrievals Conference left somewhat of a nasty taste in my mouth. I loathe Las Vegas; it is everything that is wrong with the American dream, and I always feel unclean when I have been there for any length of time. However, as Nick and I had intended, it got me over to the United States for the week at someone else's expense, and while I was there I managed to make some important contacts, and achieve a surprising amount of field research.

However, I also achieved something else. Without meaning to, I had thrown my hat into the ring, and announced my interest in the El Yunque UFO crash of 1957. Suddenly, once again, and much against my better judgement, I had - once again - become a UFOlogist.

Over the intervening months I started receiving eMails from people from within the UFO community. Some of them were just paranoid drivel. Some wanted to know why someone like me - a self-confessed UFOlogical apostasist - was daring to re-enter the field. Some, however, sent me genuine items of information, which have been invaluable as I prepare this present volume.

Six months after leaving Las Vegas I was back in Puerto Rico. Nick and I were driving along the twisted mountain roads that lead up from Canóvanas into the mountainsides.

Trying to explain a tropical afternoon to someone who has never been there is an almost impossible task. It may be because this was where I grew up as a child, but even in the hottest weather I thrive in the tropics. A hot, sweltering and humid English summer day in mid-July can knock me for six, but for some reason, even though on paper at least, the weather is far less hospitable, mid-day on a Puerto Rican July is my idea of heaven. We drove through the dusty banana groves, the huge grey-green leaves flapping like elephants' ears in the tiny breeze; the tiny little bananaquits - sparrow-like birds, like psychedelicised canaries (and the 'yellow bird' of the eponymous calypso song) - flitted from tree to tree chittering wildly. The little grey and brown ground-doves scampered across the leaf litter beneath the trees, making mournful hooting sounds, quite out of kilter for such a small and unassuming bird. Black, red and yellow tropical butterflies drifted lazily across the road in front of us, and as we meandered higher into the mountains and the air got cooler every little house seemed to have an enormous flame-of-the-forest tree standing guardian over it in the front of the garden. Whether these were the same flame trees which originated in East Africa and spread across the British Empire I don't know, but they certainly looked the same, and despite the insistent pounding of the hardcore punk rock which came out of the CD player on the dashboard, and Nick Redfern continually complaining about the eccentric road signs in broad Brummie, it was easy to believe that I was once again seven years old and back on Lugard Road on Hong Kong Island with my mother and little brother.

Suddenly, the landscape changed. Gone were the banana trees, gone were the trappings of civilisation, and we were now back in the wilderness; the montagne scrubland which lies between the grassland plateau and the true rainforest which sits atop the Puerto Rican mountains like a badly fitting wig. The air was cool and dry, and full of bird song. The walkie-talkie crackled. It was Carola.

"We are here boys", she squeaked enthusiastically. *"Take the next turning on the left"*. We did as we were bid, and found ourselves facing a huge, and somewhat sinister, pair of wrought iron gates. In the car in front, Carola leaned out of the driver's window and gabbled something incomprehensible into the intercom. It must have worked, because the gates creaked open and ushered us inside. After a short and steep driveway we found ourselves outside a house far grander and more opulent than anything I had seen before on the island.

It was a reasonably sized, split-level building, which had been constructed into the hillside. It had white stucco walls and a pointed, almost conical, roof that made it look, for all the world, like the house in Finland where Moomintroll and his family are said to live. There was a huge semi-circular veranda on the top level, and as we disembarked from our various vehicles, we could see a plump jolly lady of uncertain years waving invitingly to us amidst a veritable sea of exotic looking potted plants. I did a double take. She looked ridiculously like my Auntie Anne - my father's sister who had died about fifteen years before. Her name was Norka, and we had come to interview her about a series of encounters that she had experienced with the grotesque goatsucker.

One of the most irritating parts of filming is that one spends nearly as much time doing establishing shots which place whatever it is we are making the film about into contextual integrity, as you do actually filming anything. However, on this occasion the guardian spirits of cryptoinvestigative methodology were on our side; the huge articulated lorry which had been following us all week was stuck somewhere down in the lowlands, and it appeared that - for once - we had plenty of time to actually interview a witness without being impeded by the dull and repetitive business of making a TV documentary.

It soon turned out, however, that whatever it was that Norka had seen, and experienced on a number of occasions, it wasn't the chupacabra.

She told us how, about ten years before, she had found that one of her dogs was missing. After sending out a search party, she was horrified when they returned with the corpse of her pet. It was nothing she had ever seen before. All the flesh and the bones had been removed and in her broken English, with plenty of hand gestures, she communicated the sheer horror of the event to us. Apparently, neighbours of hers started reporting similar instances and when, one evening in early summer, she was driving home from Canóvanas, and saw a strange creature in the road before her, she had no doubt that this was what had caused the death of so many pets.

However, the thing she described was nothing like the chupacabra that we had grown to know so well, but it was something that Nick and I both immediately recognised. As she told us of the bipedal creature with claw-like feet and red glowing eyes we looked at each other, and in a moment which, if it had been scripted, could have come straight out of a *Scooby Doo* cartoon we said to each other, almost immediately *"Jesus man, it's the bloody Owlman!"*

Norka

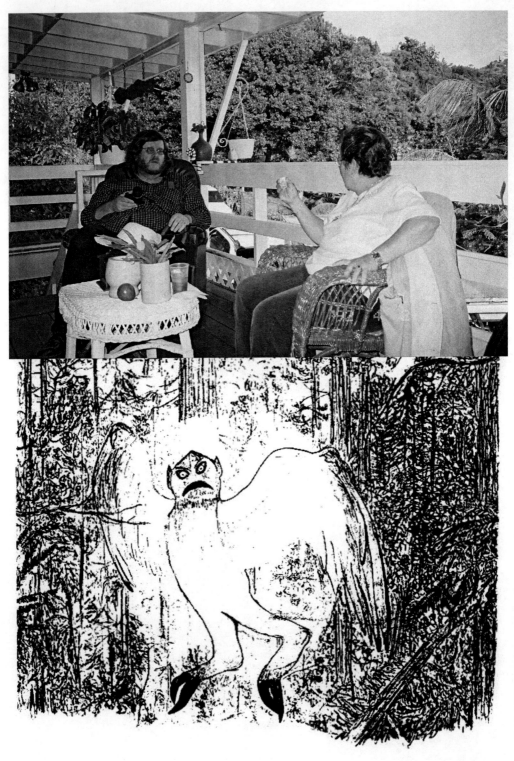

ABOVE: The author with Norka BELOW: The most iconic image of the Cornish owlman

As Norka told us of the shadowy creature with its covering of grey fur or feathers (she was not sure) my mind went back to another series of events which started off at Easter, 1976 in another monster-haunted woodland half a world away.

I owe the Owlman a heck of a lot. If it had not been for my ten-year quest to solve the mystery of the creature that has been seen again and again in the woods surrounding Mawnan Old Church in southern Cornwall, I would certainly not be in the position I am today. My book, *The Owlman and Others*, has sold more than all my other books put together and pretty well made my reputation. These days, I prefer to hunt flesh and blood animals, and do my best to stay clear of the less tangible denizens of this planet, knowing full well that hunting such things is no good for your spiritual, mental or physical health. So I was horrified at the thought that I was, apparently, back on a path that I had sworn never to revisit. However, Norka continued and told us of a series of other sightings, both of the Owlman-type creature and of a more conventional chupacabra crossing the road in the immediate vicinity of her house. It was interesting to note that, when compared to the Owman-type creature, the spiky animal seemed far more matter of fact and, although with hindsight it is easy to reinterpret events to suit one's current mindset, I am sure that, whereas Norka was only too aware that the Owlman was no flesh and blood animal, the spiky animal most definitely was.

On a later trip to the island Nick encountered stories of another Owlman-type entity. In an appendix to the latest edition of *The Owlman and Others* he wrote:

"In September 2005, I travelled to the island of Puerto Rico for a week, with a team from Red Star Films of Canada, to make a documentary for the nation's Space Channel on animal mutilations that was titled *Fields of Fear*. The focus of the filming was the Vampire-like Chupacabras. Certainly, most of the many cases that we investigated *did* fall into the category of the island's most famous bloodsucker. One incident in particular, however, was significantly different and seemed to eerily parallel the weird events at Mawnan Church in the hot summer of 1976.

It was five days into our excursion on the island and, after a delicious lunch, we headed for a place of worship known as the Church of the Three Kings, which was a wonderful building that looked like a combination of a Medieval English castle and something you would see in a dusty Mexican town in one of those old spaghetti-westerns. But this structure was surrounded by dense, green woods on all sides rather than desert, and was topped off with a large, five-pointed star that reached out to the heavens from the church's flat roof. Our interviewee was a barrel-chested figure named Pucho who had an amazing tale to tell us.

As the crew set up the equipment, little kids, excitedly chattering, came running out of the woods to see what all of the fuss was about, and a teenager insisted on riding by at every moment on a noisy, rusty old motorbike that was definitely well past its prime. Pucho shouted something in the direction of the bike and the rider, looking hurt, vanished into the sunset, never to be seen again. Pucho then settled back, folded his arms, and proceeded to relate his story.

It was long after the sun had set on a weeknight in February 2005, said Pucho, and he was walking past the Church of the Three Kings, when he heard what he described as a "loud roar" coming from a particularly dense section of trees adjacent to the building. And, as Pucho elaborated, he was amazed to see a huge, feathery bird come looming out of the tree canopy. Oddly, the bird did not merely fly out, but seemed to levitate vertically in the fashion of a helicopter, before it soared high into the air and headed off in the direction of a nearby farm. Stressing that he had never before or since seen such an immense beast, Pucho could only watch in stark terror and utter shock until the creature was out of sight. Notably, several days later the farm where the creature was seen flying towards suffered a number of horrifying attacks on its livestock, and of a type that Count Dracula himself would have been proud.

I was immediately struck by the similarities between Pucho's encounter and the three-decades-old events in the woods of darkest Cornwall. One of the witnesses to his Owliness, Jane Greenwood, described her sighting in the summer of 1976 to the *Falmouth Packet* newspaper, after a number of earlier reports had made the headlines:

"I am on holiday in Cornwall with my sister and our mother. I, too, have seen a big bird-thing. It was Sunday morning, and the place was in the trees near Mawnan Church, above the rocky beach. It was in the trees standing like a full-grown man, but the legs bent backwards like a bird's. It saw us, and quickly jumped up and rose straight up through the trees." And as Jane had perceptively asked: "How could it rise up like that?"

How, indeed? It was *exactly* that question that Pucho wanted an answer to twenty-nine years later. Personally, I wanted to know what sort of creature it was that seemed to manifest in woods adjacent to sacred grounds on opposite sides of the world and decades apart, and where the witnesses described the appearance and movements of the animal in practically an identical fashion? It is a question I *still* ask myself – obsessed, as I am, with such diabolical weirdness.

Pucho was very matter-of-fact about the encounter, cared not a bit about publicity, and just wanted to know what was afoot in darkest Puerto Rico. None of us could help him, beyond confirming that the island was a truly strange place; and whatever it was that he saw, it had no place roaming around the wilder parts of the island. He laughed loud as if to say: "*You* are telling *me*?" On that note, we said our farewells. But of one thing I was certain: whatever the Owlman was, the creature was not restricting its activities to the southwest of England. The beast – be it a physical reality or some form of diabolical thought-form – was on the move".

Much to my embarrassment, Norka then started asking me about UFOs. Nick is far more famous in this field than I am, and has written a number of best-selling books on the subject, whereas I have only tipped my toe into the UFOlogical ocean when I wanted to make a quid or two. However, it appeared that Norka was quite interested in flying saucery and had read some of the accounts that had appeared on the Internet following my ill-starred appearance in Las Vegas the previous November. I asked her whether she had heard about the 1957 crash, and to my great delight, she told me that not only had she heard about it, but that her family had owned the land on which it had happened. She confirmed Reuben's story to the letter, and added details of her own. It was true that shortly after the Government had leased the land for something like fifteen years, and it was also true that covert military operations *had* taken place on that part of the mountainside for many years, both before and after.

David's cell-phone rang. It was "Cheese". He told us that the articulated lorry was now about half a mile away, and so - almost as an afterthought - as we were preparing to go through the whole laborious process of filming this segment of the TV show, I asked Norka whom I would have to approach to go back to the crash site legally. After all, I explained, on my previous visit we had been filming without a licence, and I was pretty convinced that we had strayed into the militarised zone without realising it.

"Don't be silly", she smiled. *"You need look no further. I own it. You can go when you like."*

The land had reverted to her family some time before - at least a decade, put precisely when she wasn't sure, and she would be happy to let us do anything we wished to there.

I was recovering from that blow, when I heard the unmistakable grinding gears of the articulated lorry limping up the steep hillside, and knew that it was time to go to work for a living.

An hour or two later we had a chance to continue our conversation. There were several other things that I wanted to know about the wildlife of the area, but first I asked whether she knew what exactly the American Government had done on her hillside. She didn't, but another woman - a few years younger - who had been sitting quietly in the shadows all afternoon piped up that she had heard that this was where the Americans had tested Agent Orange. That would explain why I had discovered so little wildlife there, I thought to myself, and changed the subject quickly.

Over the years, I had picked up quite a few stories about giant snakes in the jungles of Puerto Rico. Norka knew what I was talking about, but believed that they were merely exotic pets that had grown too large, and had been released. Sadly, she had not heard any stories of crested snakes. Although there are no true crested snakes known to exist today, rumours of such things can be found across the tropics, from the West Indies to Australia, and from South-east Asia to Africa. Many zoologists, including my friend and colleague Richard Freeman, believe that they are surviving examples of a family called Madtsoids – animals which could reach immense sizes, and which have been supposedly extinct for millennia.

I also asked her if she could shed any light on the enduring stories that a colony of chimpanzees had been accidentally released into the jungle about fifty years before. She confirmed that she had heard the stories as well, but told us that she believed them to have been exaggerated. Some rhesus monkeys were known to have escaped from an immunological laboratory near Ponce in the 1960s, and as far as she knew, their descendants were still there. But they had never become numerous, and certainly had never strayed anywhere near the El Yunque rainforest.

I was disappointed with this. For the preceding forty-eight hours I had been niggled by the story we had heard at the ranch next to the cock-fighting stadium. The farmer had been adamant that the cage in which his chickens had been kept, had been opened. Having examined the catch, I could see that it would have been nigh on impossible to have opened it unless you had opposable thumbs - like a man, a monkey or an ape. Having also heard that a wiry hair sample had been left behind in the mesh, I had convinced myself that this incident, at least, was probably the work of a monkey or ape. Could Norka have been wrong?

She was so sensible and matter of fact when describing the natural history of the area, that I was minded to take what she told us about the 1957 UFO crash at face value. Unprompted, she returned to the subject. She believes that there had been several other crashes in the mountains over the years. She had often wondered, she said, whether the creature she had seen that had so irresistibly reminded me of the Cornish Owlman was somehow linked with these crashes, and that during the past few years she had become convinced that this creature, whatever it was, was trying to contact her. My hackles would have risen at this point, had not something extraordinary happened. Norka's friend, who again had said nothing during the intervening hours, pointed at the dense forest just above us. There, in full view, flying slowly across a gap in the trees was a Puerto Rican parrot; one of the ten rarest birds in the world, and something which I never imagined I would see properly in the wild, let alone twice in as many days. Truly, these forests are a place where normal logic does not really apply.

Sadly, it was time to leave. Our hostess was in failing health, and she was obviously tired. As we drove in convoy down the twisty hill roads towards Canóvanas, Nick and I mused on what we had been told. The conspiracy theories surrounding the American military takeover of much of the El Yunque were many and varied. Over the years, we had both been told of covert military operations, UFO crash retrievals, genetic experimentation, and even pitched battles between humans and aliens. Now, from an unbiased source - someone who is not only interested in the subject of unidentified flying objects, but is also the lady who owned that part of the jungle, we had had it confirmed that the American Government has certainly tested pesticides there, and had carried out jungle warfare training for troops who would be sent to Vietnam, but that there was no real evidence that anything more sinister had taken place.

It is certain that a lot of the stories linking the American Government's activities on the island with extraterrestrial intelligences have come about as a result of the Arecibo Observatory. Built in the late 1950s, it is one of the most technically advanced observatories in the world and for many years has been linked with SETI (Search for Extra-terrestrial Intelligence); certainly the best known attempt by conventional scientists to find out whether we are, indeed, alone in the universe.

The construction of the Arecibo telescope was initiated by Professor of Cornell University, who originally intended to use it for the study of Earth's ionosphere. Originally, a fixed parabolic reflector was envisioned, pointing in a fixed direction with a 150 m (500 ft) tower to hold equipment at the focus. This design would have had a very limited use for other potential areas of research, such as planetary science and radio astronomy, which require the ability to point at different positions in the sky and to track those positions for an extended period as Earth rotates. Ward Low of the Advanced Research Projects Agency (ARPA) pointed out this flaw, and put Gordon in touch with the Air Force Cambridge Research Laboratory (AFCRL) in Boston, Massachusetts where a group headed by Phil Blacksmith was working on spherical reflectors and another group was studying the propagation of radio waves in and through the upper atmosphere. Cornell University proposed the project to ARPA in the summer of 1958 and a contract was signed between the AFCRL and the University in November of 1959. Construction began in the summer of 1960, with the official opening taking place on November 1, 1963. [1]

The telescope has undergone several significant upgrades over its lifespan. The first major upgrade was in 1974 when a high precision surface was added for the current reflector. In 1997 a ground screen was installed around the perimeter to shield from ground radiation and a more powerful transmitter was installed.

The Arecibo telescope has made many significant scientific discoveries. On 7 April 1964, shortly after its inauguration, Gordon H. Pettengill's team used it to determine that the rotation rate of Mercury was not 88 days, as previously thought, but only 59 days. In 1974 Hulse and Taylor discovered the first binary pulsar PSR B1913+16.

In August of 1989, the observatory directly imaged an asteroid for the first time in history: asteroid 4769 Castalia. The following year, Polish astronomer Aleksander Wolszczan made the discovery of pulsar PSR B1257+12, which later led him to discover its three orbiting planets (and a possible comet). These were the first extra-solar planets ever discovered.

SETI, which relies greatly on the Arecibo Observatory, is an attempt to create a communications system using a powerful transmitter and a sensitive receiver, and use it to search the sky for extraterrestrial worlds whose citizens have a similar inclination as terrestrials. A basic assumption of SETI is that of "Mediocrity": the idea that humanity is not exotic in the Cosmos but in a sense "typical" or "medium" when compared with other intelligent species. This would mean that humanity has enough similarities with other intelligent beings and therefore communications would be mutually desirable and understandable. If this basic assumption of Mediocrity is correct, and other intelligent species are present in any number in the galaxy at our technological level or above, then communications between the two worlds should be inevitable.

SETI is still no trivial task. The Milky Way galaxy is 100,000 light years across, and contains a hundred billion stars. Searching the entire sky for some far-away and faint signal is an exhausting exercise. The SETI program gained fame on August 15, 1977 when Jerry Ehman, a project volunteer, witnessed a startlingly strong signal received by the telescope. He quickly circled the indication on a printout and scribbled the phrase "Wow!" in the margin.

This signal, dubbed the Wow! signal, is considered by some to be the most likely candidate from an artificial, extraterrestrial source ever discovered. However, the signal has never been repeated, and the source has never been confirmed. But it has gained a fair amount of notoriety.

---

1. A good start to reading more about SETI, and indeed the Arecibo Observatory is McConnell, Brian; Chuck Toporek (2001). *Beyond Contact: A Guide to SETI and Communicating with Alien Civilizations.* O'Reilly. ISBN 0-596-00037-5.

Still in convoy, we drove back to Canóvanas, and found ourselves parked outside a ramshackle wooden building with a corrugated iron roof. I had been there before. It was a café bar where six years before, Graham and I had sat eating plates of corned beef hash and watching the world go by. Sitting in the corner, with a broad grin splitting his face from ear to ear, was Ismael, accompanied by a surly looking bloke sporting a comedy Zapata moustache, puffing earnestly away on a bedraggled looking cigar. Someone - I forget who - purchased beers all round, and the entire crew - eleven or twelve of us, including the drivers, and various small boys who clutched paper cups full of virulent coloured pop - sat down to discuss our next move.

It turned out that Ismael's companion was a well-known chupacabra witness. Apparently, he had some photographs, and video footage of alleged chupacabra footprints that he wanted us to see. However, there were two problems:

Firstly, he wanted five thousand bucks even to let us look at them. Despite Ismael's entreatments, he would not be budged, and I felt increasingly sorry for Ismael. My new friend was obviously highly embarrassed by the developing situation. The two locos gringos had actually turned out to be quite nice, and he was doing his best to uphold his position within the chupacabra-orientated universe. After all, apart from - possibly - the good Mayor Soto, as far as chupacabra hunting around the world was concerned, Ismael was th'gaffer. He was finding, to his cost, what I had discovered many years ago – namely, not to allow your dodgier friends anywhere near your professional life. This unprepossessing looking fellow was obviously a good mate of Ismael's, but his intransigence was obviously causing the poor chap no little embarrassment.

The second reason that I took an - if not quite an instant, then pretty soon after - dislike to the insalubrious would-be vendor of (what I was rapidly beginning to suspect were) dodgy chupacabra prints, was that he looked frighteningly like an exceedingly second-hand car dealer, and part-time chef, that I had had the misfortune to know in Exeter some four years previously.

It had all started when I bought an opulent looking Rover from a small ad in one of the yellow free pages. The bloody thing packed up within weeks, and I should have been warned. Even now I am not sure why, but when I went round to the car salesman's house to demand my money back, somehow I ended up in a business partnership with him selling second-hand cars to the impoverished underclass of southern Exeter. It was probably because I knew that this was the only way I was going to get my money back, but the idea of getting a foot several rungs into the desirable automotive ladder (much the same as the property ladder), but noisier, less reliable, and with worse dress sense) had a certain low-life appeal to me, and I commenced my brief career as a used car salesman.

My new partner, who had a surprisingly aristocratic name for an obvious oik, basically took me to the proverbial cleaners, and I ended up several grand out of pocket and with no form of transport. The sticky end came when, after retrieving one of my cars from outside his abode, I was told that an associate of his called 'Big Colin' had left a computer in the boot, and that as said computer had now disappeared into the ether as a dream disappears at the break of day, 'Big Colin', and his friend (an imaginatively named 'Little Colin') were "coming round my place" to remove my external genitals with a claw hammer and blowtorch.

My old mother used to tell me that I should never judge people on first appearances, but basically whenever I have done just *that*, things have usually worked out OK for me. I suppose that is perfectly possible that Ismael's mate (whose name I forget, if we were, indeed, ever introduced) was a fine upstanding young chap. He may have been a churchwarden, a pillar of the local youth club, a tireless community worker who was always kind to animals, and who never forgot to send his mum a box of chocolates on

Mother's Day, but somehow I doubt it. Anyway, we didn't have five grand to spare, so we politely declined his offer, made plans with Ismael to rendezvous at the same venue the next day, and we drove off towards San Juan.

As I am sure that I have already mentioned in this narrative, I have certainly said on other occasions, making films can be a pain in the arse. Much of what you see on television is pure artifice and bears very little relation to what actually happened. To give David his credit, compared with other directors with whom I have worked he asked a minimum of fannying about from us, but we still had to accomplish a certain amount, and so during our journey back to San Juan we had to film innumerable short sequences that could be used as linking shots, showing Nick and me driving intrepidly up and down the highways and byways of the island.

I have been doing this for years, and a long time ago, I discovered that the best way to while away the long hours spent driving up and down the same expanse of highway again and again and again (and again) was to play silly buggers with the walkie-talkies. Not for the first time on this expedition I began to sorely miss my old friend and colleague Richard Freeman. Together, we have whiled away the duller parts of many film and TV projects, and indeed expeditions by playing a singular game of his invention, called "Find the Fish". To anybody without a ground in ichthyology, this is a pointless, and incredibly tedious exercise. It is basically the old parlour game of twenty questions, except with fish. One person starts the game off by saying *"I've got a fish!"* The usual repost (although this is not compulsory) is *"Is it an elasmobranch?"* (meaning: has it a cartilaginous rather than a bony skeleton?) At this point everyone else within earshot usually goes to sleep out of sheer disgust, but Richard and I have had hours of fun with this innocent pastime. Whether the pleasure we gain is from impressing the other with our mildly encyclopaedic knowledge of denizens of the finny-tribe, or whether we gain a perverse satisfaction from irritating our companions, I am not prepared to admit, but on this particular July afternoon, driving through Puerto Rico there were no balding Goths in site, and so Nick and I were forced to fall back upon our resources for entertainment. We played every stupid kids car game you can think of, amusing ourselves by teaching the Spanish-speaking crew to play `I Spy`, and swapping increasingly inane and scatological jokes with David and Kevin in the car ahead.

Then, almost without warning, we pulled over into the side of the road, and the day's actual work resumed.

For me, at least, the first part of this filming was simple. To start off, I had to drive with Carola in her sumptuous air-conditioned people-carrier, as David travelled with Nick and busied himself by trying to find a stretch of road that was marginally less unattractive than the others, we drove along, and Carola started to tell me about what the chupacabra actually meant to the people of the Island of Paradise.

She believed in it, of course. But then again it seemed that everyone on the island believed in it, if not as a physical creature with an objective existence of its own, as a symbolic entity.

For years I have noted that a lot of people tell me that *"They believe in the Loch Ness Monster"* when, in fact they believe in no such thing. They have very rarely actually examined the evidence – for, or against - in much detail. When they say that, what they *really* mean is that they consider themselves to be open-minded, and not prepared to bow down to the rules of scientific or cultural orthodoxy.

So, at first, this is what I thought that people like Carola and "Cheese" meant, when they told me that they believed in the chupacabra. However I soon began to realise that I was wrong. It is completely pointless to try and judge an alien culture by the mores of your own, and Puerto Rico is very much an alien culture. It is a land where the gulf between the living and the dead, and between beings of flesh and blood, and beings of the aether - at least in the minds of those who live there - is by no means as pronounced as it is in other parts of the world.

The dead are dead, sure, but the belief in ghosts, and the free-and-easy communication between the living, and the spirits of those who have passed on is universal. Every village and town has a plethora of ghost stories. The township of Aguirre, for example, boasts the following spectres (and this is only from a cursory peep at s local ghostwatch website): [1]

- A headless woman was observed hitch hiking along a dark road outside Aguirre. The ghost did not care that there was someone else there.
- The ghost of a coal miner can often be seen drinking gasoline from a pump at a gas station in Aguirre.
- The ghost of a young Indian warrior has been witnessed on numerous occasions mailing a letter at an Aguirre post office.
- The ghost of a badly mangled hunter dragging a dead bear can often be seen seated at a table in an Aguirre house.
- A man with no head can be witnessed very often going through the fridge in the kitchen of an Aguirre home in the early morning hours.
- The ghost of a young woman with a rope around her neck can often be observed looking at people in an Aguirre home through an air vent.
- A female shape has often been witnessed by a man camping at a campground outside Aguirre.
- The ghost of a badly burned woman can often be observed picking flowers in the front yard of a house in Aguirre
- The ghost of a farmer in a straw hat was observed in a deserted area outside Aguirre. The eyewitness was frightened and ran away.
- The ghost of a young man wearing a confederate uniform is often observed smoking a pipe under a streetlight in Aguirre.
- A black cat that turned into a woman has often been witnessed sitting in a chair in a house in Aguirre.
- The ghost of an engine driver can be seen very often standing in the middle of the road outside Aguirre.
- The ghost of a woman with a bag tied around her head has been observed on a few occasions walking from house to house in the early morning hours before sunrise on an Aguirre street.

Some of these stories are particularly perplexing. The young soldier in the uniform of the Confederate Army, for example, makes no historical sense. During the American Civil War, Puerto Rico was still – nominally, at least - a province of Spain, and took no part in the conflict on either side.

As described elsewhere in this book, the belief in UFOs is similarly almost universal, and therefore, the fact that Carola was a believer in the chupacabra was no real surprise. She also believed in UFOs, ghosts and fairies, and had experienced a few strange things of her own.

However, just as she was about to tell me of her strangest experience, the walkie-talkies crackled, and David's voice came over the ether, asking if either of us knew any of the songs about the chupacabra. Whereas once, I would have been surprised by this question, now I was an old hand on the subject, and quoted the lyrics of one particularly crass ditty:

---

1. http://www.ghostsofamerica.com/

eh, Chupacabraaaaaa ..

Some people say they have seen the Chupacabra,
after some tequila, he was dancing the lambada,
he doesn't know where he has come from,
making all of Mexicana's looking like a Dum-Dum,
they say that he's a monkey,
I think the whole story is just a little funky.

There's a buggy eyed creature that they call the Chupacabra,

It's only silly, but you see we really gotcha,
like hearing the lambada or the hooky cucaracia,
we started a sensation,
about a creature that is just hallucination,
we did it for the money,
but we duped you and we think it's really funny.

David snorted in disgust. He had obviously been hoping for an ancient Latino folk ditty, which he could use to cinematic effect in a similar way to how Ken Russell had used a version of *The Ballad of the Lambton Wurm* in his delightfully nutty retelling of Bram Stoker's, *Lair of the White Worm*. You could hear the disappointment etched into his vocal chords. He said something terse and signed off.

The search for a suitable location continued. Carola and I resumed our interrupted conversation.

She told me how, a few years earlier when she had been in her late teens, she had been out for a picnic with some of her friends. I don't know whether it was just my testosterone fuelled imagination, or just the delightfully sultry way in which our Field Producer managed to imbue every sentence with a miasma of sensual overtones, but the implication was that this would have been a more - ahem - `intimate` affair than just cheese and onion sandwiches and a bottle of pop, but she said that her friends and she had been planning to have an alfresco meal on the shores of a beautiful lagoon where the Canóvanas River meets the sea.

As they approached the lagoon on foot, it was as if the sky `fizzled` and they were suddenly in a different reality. They saw strange lights in the sky, and although only what seemed like a few minutes passed, they were there several hours, and that none of them had ever felt comfortable talking about the experience again.

This wasn't the first `missing time` experience in Puerto Rico that I had been told about. As I wrote in *Only Fools and Goatsuckers:*

"☐there is a well attested story that a well known professor from an Ohio University disappeared for twelve days in the forest within the past couple of years, only to re-appear apparently in good health and with no memory of how he had spent the intervening week and a half. Unfortunately the "Government Research Lab" which probably does exist is undoubtedly something to do with the aforementioned pest control projects and the Professor most probably had `slipped his leash` with the intention of commiting some illicit lechery in one of the brothels that I am reliably informed exist in the less salubrious parts of Old San Juan.

My friend and colleague Scott Corales, an expert in the forteana of Puerto Rico (more of him later) wrote to me with reference this case:
"I have a lot of information on the absent minded professor, but its all in Spanish! The case never really held my attention so I never translated anything on it. The verdict on the case was highly unfavorable (crude remarks were made about the fact that the

was staying a hotel that caters to a gay crowd) and the professor refuses to discuss the matter any further. The case was referred to me by UFOSCIPR, since I'm only 8 hours away from Bowling Green Universtiy, but I passed it on to MORA (Mid Ohio Research Assoc.) and even they were unable to get any information out of Darby!"

However, Carola's story was so obviously heartfelt and genuine, that it makes me wonder - slightly guiltily - if I should have been a little more understanding about the story of the poor lost professor from Ohio.

Eventually, after about an hour, David found what he considered to be the ideal location for the next scene of his burgeoning masterpiece. He went back to his own car clutching the walkie-talkie, while the camera crew (who were beginning to enjoy all these high jinks immensely), sat precariously in what had been my seat, the soundman - "Cheese" - positioned so his not inconsiderable posterior was waving precariously over the side of the open-topped Jeep. Nick was instructed to talk to me as if I was there, sitting next to him.

However, David (being an Oscar-winning Director, after all), was not satisfied. Nicholas Redfern is a damn good author, and an all round good egg, but he is a lousy actor, and no matter what he said, it was obvious that I was nowhere to be seen and that he appeared to be exhibiting the first signs of a personality disorder, by talking to himself. In order to add verisimilitude to an otherwise unconvincing scenario, I was instructed to carry out a conversation with Sir Nicholas of Redfernshire via the magic of walkie-talkies. Ernesto - a small boy who was somehow attached to one of the crew - was instructed to crouch in the footwell in front of the seat, holding the walkie-talkie in his sweaty hands, and relaying whatever pearls of wisdom Redfern would impart.

There was a third walkie-talkie, in the breast pocket of the jacket worn by our beloved director.

"Um, David dear boy," I began, diffidently. *"What the hell do you want us to talk about?"*

This question seemed to throw him. He hadn't thought of that.

"Do you want me to reprise some of the things we were talking about earlier?" I asked, but after a short gap he said no. He wanted to wait until he could see us talk to each other about matters of importance within the parameters of the expedition, but he wanted to establish that Redders and I were "buddies" (which of course we are).

"What about giving us an example of your famous British humour?" he asked, whereupon Nicholas and I embarked upon a tirade of sexism and smut from the pages of *Viz* comic that would have hardly been recognisable to anyone who was not either one of the cognoscenti, or one of the less salubrious inhabitants of a Tyneside council estate.

As Nick and I regaled anyone who happened to be listening over the airways to an ongoing saga featuring Ma and Pa Bacon, their son Biffa, and Cedric the Puff, there was a stunned silence from the directorial car.

"Fither, I'm off out to clobber some bastard in the mush", Redfern said in a peculiar mixture of his native Brummy accent, his acquired Texas twang, and the Geordie patois that he was only vaguely successful in pulling off.

"Give the bastard one from me kidder", I would reply in an only marginally less convincing Tyneside accent.

"He called wor lass a heemersex", Nick replied and then muttered something about me spilling his pint. As we drove along the dusty tropical highway we passed a large and scabby dog urinating against the wall of a house *"better shoot the bastard, before it homs up a bairn",* I grunted utilising what was possibly the worst quasi-Geordie enunciation in the history of the human race.

The walkie-talkie crackled into life, and finally David's exasperated tones burst through the tropical afternoon. *"What the hell are you two limey idiots talking about?"* he asked with genuine pain in his voice. *"Anyway, pull over at the truck stop down here and we'll swap you over. It is time to film you from behind."*

Both Redfern and I collapsed into hopeless giggles, and came out with childish sexual slurs at poor David which would, I am sure, make my wife blush as she sits here typing this record of her husband's idiotic behaviour on the other side of the world.

We did as we were told, and pulled in. As we did so, a tiny pick-up truck loaded so high with bananas, that the teetering mound of yellow fruit appeared to be always in imminent danger of coming to grief, drove passed. Its radio was blaring, and both the passenger and the driver's voices were raised in mildly harmonious song. *"Can you two idiots sing?"* asked David, trying to suppress a grin. Redfern grunted, and I was about to deny having any vocal talent whatsoever when David looked straight at me and said:

"There's no use denying it, I know you used to be a rock singer......", he said accusingly.

"Yeah, but it didn't mean I could sing", I said. *"How about you and Nicholas regaling us with some of your old English folk songs?"* he suggested, and Nick and I looked at each other and grimaced.

"Er, we'll see what we can do", said Nick, and we got back in the Jeep, as the assorted crew; Kevin the cameraman, "Cheese" and Ernesto, another bloke called Ramon (who I strongly suspect was just some mate of the crew who wanted a free lift back to San Juan, and had absolutely nothing to do with the more serious business of film-making, or indeed chupacabra hunting), and a small boy who seemed to be a mate of Ernesto, crammed into the back of the Jeep to capture our progress back towards the island's capital.

By this time we were becoming close friends with the crew, and their motley collection of hangers-on, and after I made a valiant attempt to accede to David's wishes by singing a verse or two of *Widecombe Fair* to the accompaniment of cat calls and rude remarks from the Brummy ingrate sitting next to me, the attempts at folk singing soon petered out.

"Cheese" began to tell us about the folk music of Puerto Rico, and promised to bring me a CD of some local songs the following day. Ernesto and the other small boy were coaxed into singing some of the folk songs of the island, and the rest of the crew soon joined in. They were catchy tunes, and although I had no idea what we were singing about, I joined in the choruses with gusto. Even Redders bobbed his head rhythmically in time with the music, and a splendid time was had by all.

After one particularly bawdy sounding number, I asked "Cheese" and his compadres what it was all about. His friend (whose name I still don't know) told me in faltering English:

"Eet ees, how you say, when the men from one village go to another village, to, how you say, make-a jig-a-jig with the ladies", he blushed and everybody laughed uproariously. *"You have same songs like that in Angleterra?"* he asked with a saucy smile, and both Redders and I had to admit that, yes, of course we did.

"You sing thees song now", he laughed, and the assembled company, including the small boys started to urge us to enrich their lives with a piece of cross-cultural pollination. So Redfern obliged:

*"Walsall bootboys we are here,
To shag your women and drink your beer"*

he carolled, as I attempted, in my execrable Spanish, to explain the finer points of the traditional courtship ballad of the discerning West Midlands football hooligan. In this happy, and relatively harmless manner, we drove into the suburbs of San Juan, where we dropped-off our passengers. We then made our way towards the hotel, ordered delicious omelettes at the bar, washed them down with several bottles of beer and a strawberry daiquiri, and retired to our respective rooms to sleep.

The Margarita Diary,
Chapter 6 Comments:

Back in the mid-to-late 1990s when, in the popularity stakes at least, Mulder and Scully were the Posh and Becks of their day, and the British newsstands were saturated with magazines on all-things ufological, I spent a lot of time writing books and articles on UFOs. Back then, I freely confess, I was a paid-member of the 'I want to believe'-brigade.

Well, I still want to believe to this day. Unfortunately, however, I'm not able to; as much of what I accepted as fact back then, I simply don't now. Whether due to the passage of time, the fact that many of my views were turned upside down by new discoveries and revelations, or more likely a combination of both, aside from music, clothes, beer and several other things, the Nick Redfern of 1997 was not the same Nick Redfern who travelled to Puerto Rico in 2004 with Jon.

I was already deeply involved in the writing of my *Body Snatchers in the Desert* book at the time of our trip, and my views on 'UFO Crash-Retrievals' had already radically changed. And so, I cast a weary and wary eye in Jon's direction when I first heard tell of the 1957 UFO crash deep within the heart of the El Yunque rainforest.

Yet, I most certainly was not a sceptic; and so listening to the data Jon had uncovered was fascinating - even more so, if the event had indeed somehow had a radical effect on the life-forms that inhabited the area of the crash - both human and animal.

And that's why I have such good memories of our meeting with Norka, too. Not only was she able to fill in some of the gaps suggesting that at least *something* had genuinely crashed on Puerto Rico back in the 1950s, but she was also a veritable fountain of knowledge on all-things monstrous too. As long as I live, I will never forget that moment when Norka told us of one of her encounters and Jon and I turned to each other and realised that the beast Norka had seen was practically identical to the notorious Owlman of England - a creature that Jon had hunted, and been haunted by, for years. It was truly a pivotal moment in that memorable week.

As we sat on the balcony of Norka's beautiful home high in the hills of El Yunque, sipping cold drinks, listening to her stories, and with the sun bathing down on us, I knew that we were experiencing something very special, and that beneath its beautiful exterior, something - or *some things* - dark, ominous, dangerous and bizarre dwelled on the island. **NR**

Part Three
GENESIS

*"We sail tonight for Singapore,
take your blankets from the floor
Wash your mouth out by the door,
the whole town's made of iron ore
Every witness turns to steam,
they all become Italian dreams
Fill your pockets up with earth,
get yourself a dollar's worth
Away boys, away boys, heave away*

*The captain is a one-armed dwarf,
he's throwing dice along the wharf
In the land of the blind
the one-eyed man is king, so take this ring"*

Tom Waits/Kathleen Brennan `Singapore`

Chapter Seven
Hasta la Vista baby

One of the things that I like most of all about being on an expedition, is that it gives me the luxury of being able to concentrate on the job at hand, and whatever portions of my psyche are left over can concentrate on the pleasures of life like the aforementioned breakfasts. As the CFZ has become more and more successful, I find myself in the unenviable position of always having some pressing matter to which I must attend. Whereas, back in the halcyon days of the early 90s, we were just a gang of well meaning enthusiasts who put out a magazine now and then, and never numbered more than a hundred, by 2004, in our twelfth year of existence, we had somewhere between three and four hundred members, a punishing publication schedule, several foreign expeditions a year, and plans were well under way towards starting a full time visitors' centre and museum.

Back at home in Exeter, the days when I would write a couple of articles, put the answerphone on, and then slope down the pub for the rest of the day, had long gone. I usually found myself working 8 to 10 hour days seven days a week, and as I lived above the shop with my friend and colleague Richard, even our leisure time was often spent making plans for expeditions, and investigations that we couldn't possibly afford.

It's ironic, therefore, that even though I was thousands of miles from home, in the sunny Caribbean, I had more time to myself than I had been used to for some considerable time, and so; arising early, a bracing shower, and then a long leisurely breakfast followed by a couple of equally long and leisurely cigarettes, and a couple of chapters of derring-do from my trusty volume of *Bulldog Drummond*, was sheer and unaccustomed luxury. I am not normally known as an early riser, but with the prospect of a couple of hours of lazy luxury before a long and hard day in the field, I tend - when in the field - to get up at an hour which would astound any of my friends, loved-ones, or associates back in the UK.

I was drinking my third cup of coffee, smoking my second cigarette, and gasping at the despicable exploits of the surly, hooked-nosed conspirators who were plotting their evil deeds in the East-End of London in the mid-1920s, and there was a tap on my shoulder. I turned round, expecting to see the saturnine features of Nick Redfern, but was agreeably surprised to see "Cheese" beaming down at me.

"Hey dude, come and join me," I said. *"There is still some coffee in the pot"*, and he did just that.

For about half an hour we sat and talked, and "Cheese" - once again - confirmed my belief that Puerto Rico is an incredibly strange place. I told him about my discussion with Carola about the mysterious lagoon at the mouth of the Canóvanas River. He wasn't at all surprised to hear it, and he, too, told me of the strange village by the lagoon populated entirely by people with progeria-like symptoms. I hadn't realised that this was the very same lagoon where Carola and her friends had had such a strange experience. However, "Cheese" not only confirmed the story, but also had actually been to the village himself. He painted a grim picture of an increasingly beleaguered community who was shunned by their neighbours (who to a man believed they were aliens) and received little or no help from the authorities.

We talked about the chupacabra at some length, and I think it was only then that I realised how - at least to those educated denizens of the island who had made a study of the vampiric killings - the whole affair had somewhat a feel of "ancient history" about it, because although there was still intermittent sightings of the strange semi-bipedal spiky-backed beast seen shambling across ill-frequented roads in the wilder part of the Canóvanas district, and the subject was still as fascinating as ever to foreign TV companies like ours, the killings - at least those which could be firmly linked with the chupacabra - had petered out some years before. Indeed, there hadn't been a properly attested chupacabra-related killing since 1998 or 1999. The incident in the Rodriguez sector which I had investigated, and which produced the dead chicken that I had examined back in 1998 must have been one of the final chupacabra-related killings on record.

I told "Cheese" that weird episodic incidents were nothing new in the annals of forteana, and that I had been pivotally involved with one of them when I blundered into the middle of the series of BHM sightings at Bolam Lake in Northumberland early in 2003. Such episodes happen, and move on. Often never returning to the same district. Indeed, places like Mawnan Church, where a series of monster sightings seems to be part of its *genius loci,* are in the minority. I am often asked by tourists visiting Britain for the first time for a list of places of fortean importance. I am quite happy to give these, but always feel somewhat of a heel when I point out that no-one has seen a griffin in Brentford since the early 70s (if, in fact, they did then, and the whole affair was not just a hoax), no-one who isn't barking mad or has an axe to grind (or both) has seen anything more untoward than the occasional glue-sniffer or homeless tramp in Highgate Cemetery for many a long year, and the skies above Warminster, which once resembled an episode of *Star Trek* in terms of the enormous numbers of UFOs that had been sighted there, have been empty of alien visitors for nearly four decades.

However, the sightings of what I had once referred to as 'Sonic the Hedgehog on acid' continued and, the more I studied them, the more I was convinced that there was a real, *bona fide* unknown animal; lurking in the less well-travelled by-ways of Puerto Rico. I said as much to "Cheese", and the two of us waxed lyrical on the subject of the beauty of Puerto Rico's vanishing virgin countryside until there were tears in our eyes. Suddenly our revelry was shattered by a loud belch and a shout of, *"Well I'll go to the foot of our stairs, if it in't Jonny and our "Cheese". Bostin' loike".*

I am sure he does it just to annoy because he knows it teases, but Redfern becomes more and more Brummy with every waking hour spent away from hearth and home. He sat down with us, ordered coffee and a bowl of fresh fruit (a mixture that would have explosive results later in the day), and reached into the voluminous holdall, which he seemed to carry with him at all times. Out came a large chammy leather, and to my great delight he started polishing his head with it whilst recounting an interminable and scatological anecdote about three friends of his from West Bromwich who had once bought some defective marital aids at a car boot sale. It was very funny, and could not be repeated in polite society (or even within the pages of this book), but it destroyed the reverie, and "Cheese" and I completely lost the thread of our conversation.

Then the others arrived, and before we knew it, we were off on the road again for our last full day's investigations.

As we drove into Canóvanas I began to realise that for the first time I wasn't just a stranger in a strange land. I was beginning to recognise some of the landmarks and some of them felt like old friends. We spent a particularly frustrating couple of hours driving round and round the increasingly complex one-way system that operates in the back streets of the town. I had done this before, and laughingly told Nick of my first experience filming 'drive-by cutaways' in the Canóvanas back streets. It had been back in the early January of 1998, and the whole episode had been bounced on us in the last minute by a director whom, as I have said before, had the main agenda of trying to make an entertaining documentary by dint of making its two principal stars (Graham and me) look foolish. I don't know whether he had urged us to have a large and boozy lunch, implying that work was finished for the day, before bouncing this drive-by experience upon us or not, but this is what happened, and Graham and I found ourselves driving a left-hand drive car with which I was totally unfamiliar, around tiny streets that all looked the same, scattering stray dogs and chickens in our wake, while we were both half-cut. My enduring memory of that occasion was of following the road signs marked "Transito". I had no idea what they meant, but Graham was convinced that - because in the UK, at least, Ford Transit vans are the chosen vehicles by many an itinerant hippy - or rock musician - that "Transito" implied that we were on our way to the nearest free festival where - hopefully - *Hawkwind* would be playing on the main stage.

It transpired that 'Transito' actually meant a one-way traffic system, and that - by luck, rather than by judgement - we had been going the right way! As even the *traffic* cops of the Island of Paradise are armed, I was quite glad that, once again, the Gods of cryptoinvestigation had been smiling upon us.

This time Nick was driving, we were both distressingly sober, and the only impediment to our progress was the fact that Nick was laughing so hard at my story that he drove straight into a ditch. However, no damage was done, and we continued our circuitous investigation of the back streets of Canóvanas. It must have been the school holidays, because various local urchins began to recognise us after we had passed them for the third time, and each time we drove past, as well as the stray dogs and chickens scattering before us, there was an ever-growing band of grubby Puerto Rican children cheering. I am sure that the sight of two strange Englishmen - one very fat and one very bald - driving around in circles as they listened to a CD of wild, raucous, guitar-driven music must have been highly entertaining.

The reason that we spent well over two hours driving around the same little circuit of tumbledown suburbia was so that David and Kevin could get drive by shots of us from every conceivable angle. We then drove out of the town towards the foothills of El Yunque and did exactly the same thing again, so that we could get appropriate drive-by shots with the rainforest in the background. This was marginally more enjoyable, because - instead of an audience of urchins, stray dogs, and chickens - there were exotic birds and butterflies to look at. The whole process was much quicker up here, and before we knew it, it was lunchtime, and we all went in convoy down to a little café just outside Canóvanas for lunch.

The blue tropical sky is deceptive. When I went down to breakfast soon after six that morning, and together with *Bulldog Drummond* and the dusky maiden dispensing coffee, sweet smiles, and tostados, we were the only people in the universe, and the sky had been a delicate eggshell blue. The day had promised sweetness, light, and sunshine and very little else. By mid-morning, as we were driving round and round the back streets of Canóvanas to the delight of sundry dogs, hens and street children, the sky was a bright and unforgiving blue out of which the burnished sun beat down to desiccate all before it. By the end of the morning, high in the mountains of El Yunque it was beginning to rain; little spitter-spatters of the sort we were assured, as children, wouldn't last an hour because they were sunshine showers.

As we drove down the steep hills towards Canóvanas and lunch, the sky was the colour of dirty brass, and great gobbets of rain spewed out from the heavens on to the unsuspecting earth below. Our little mountain roads were soon under several inches of water, and we were driving down the middle of what was essentially fast becoming a mountain stream. I've been in tropical rainstorms in other parts of the world at various times in my childhood, but never anything on this scale. As we drove around the next corner, a great wall of water hit us and for several minutes visibility was nigh on minimal. *"Effin' 'ell, Jon"*, said Nick, with a grimace, *"It's like August Bank Holiday in Dudley"*.

Having never been to Dudley, and indeed, having very little wish to do so, I wasn't too sure what to make of this, and I made no comment. Luckily, about ten minutes before, when the rain had started, Nick and I had had the forethought to put up the soft canvas roof of our Jeep, and so we were afforded some reasonable protection from the gathering storm. However, as the rain got harder, we found ourselves in a position where driving was nigh on impossible, and were forced to pull into a non-existent lay-by at the side of the road, praying that all the other road users would be sensible enough not to try and continue their journey, and so the fact that we were impeding a certain amount of the available traffic space wouldn't matter. The other road users must have been as sensible as we had hoped, because we didn't see anybody try and get past, as we sat in the Jeep, and decided to make the best of a bad job and wait out the rainstorm.

It was then that the next problem began to make itself evident. The canvas roof of the Jeep was not particularly watertight, and the sides - and windows - were only held on by Velcro. The driving tropical rainstorm was so intense that rain soon started trickling into the interior of the car, and the footwell on the driver's side soon had several inches of water in it. Poor Nick was soaked to the skin, and I wasn't much better. But the rainstorm soon passed, and within half an hour we were back on the road again; there wasn't a cloud in the sky and apart from torrents of muddy water spewing down the middle of the road, one would not have known that anything had occurred.

Wet, bedraggled, and looking more like a troupe of slightly unsavoury drowned rats, than the elite *crème-de-la-crème* of international cryptozoology accompanied by an award-winning film director and multi-cultural crew, we trooped into the tiny restaurant, went up to the bar and ordered most of it. We were certainly there *en masse* this time. When all the members of the crew, their children and hangers-on were finally seated, there were nineteen of us. Carola was the only woman, and she basked in the fact with a sort of unconscious lazy feminity; a bit like a pedigree Persian cat at a cat show, surrounded by a bunch of slightly feral looking toms.

I don't think that the restaurant had ever had quite so many people in for lunch at once, and so taking nineteen lots of orders, and serving nineteen three-course meals (everybody, of course, wanted different things) took an inordinate length of time. However, it was cool, comfortable, and I don't think Medella has ever tasted so good as it did that afternoon, so no-one was particularly keen to get back to filming, and we were all quite happy to sit around drinking beer, and talking inconsequential nonsense to each other.

Always being a man with an eye for the main chance, I saw an opportunity that couldn't be missed, and I told the crew - who were all, of course, local - about my experience with the lamprey I had caught in the little stream on the other side of the road. Not surprisingly, none of them knew what a lamprey was, and

"Effin' 'ell, Jon", said Nick, with a grimace, *"It's like August Bank Holiday in Dudley"*.

ABOVE: Drowned rats BELOW (and over): What had once been my beautiful river

so, marshalling my few resources at bilinguilism, I gave a brief lecture on lamprey biology, illustrated by badly drawn pictures executed in pencil on paper table-napkins.

"Aaah, eet ees like anguilla?" asked Ernesto with a quizzical grin.

Guessing (correctly as it turned out) that as anguilla meant 'eel' in Latin, it would mean the same in Puerto Rican Spanish, I nodded my assent, took a swig from my bottle of Medella, and drew a reasonably competent diagram of a lamprey sucker, explaining as I did so the peculiar biology of its unique circular jaw. One of the quieter members of the crew, a delightful chap whose name I never did learn, butted in, whispering something in Spanish to Ernesto. It turned out that his English was so minimal that he was embarrassed to display his lack of it in front of me and Nick. We may have been *locos gringos* but we were their meal ticket for the week, and he felt awkward about losing face in front of us. *"He say that he not see anguilla that suck blood but when he was boy he knew ones that hold on to stone like thees ....."*, and they both did a creditable impersonation of an eel holding on to a rock in the middle of a mountain stream using modified pectoral fins as a sucker.

Before I could comment on this, the waitress, who - like waitresses the world over - had been happily eavesdropping on our conversation, butted in, and said: *"Senor, you should ask my children - they go for feesh in the river".*

The waitress went and fetched her children, and I told the story once again, and dispatched them down to the river with promises of largesse if they could catch one for me. I had to explain that they were looking for something that looked like an eel but with a circular sucking mouth. I told them that I wanted the animal alive if possible.

It has always been something that I have noticed amongst media types, that they *do* enjoy a mid-day meal.

One of the best satires on this came from the late, great Douglas Adams who told of how the editorial staff of the *Hitchhiker's Guide to the Galaxy* used to take sponsored lunch-breaks in aid of charity, in memory of one of the earlier editors who disappeared after one particularly excessive mid-day meal, never to be seen again. Indeed, he noted that all subsequent editors of the exalted guide were really just deputy-editors because despite the sign over the missing editor's chair which read, 'missing presumed fed', there were some that - even many years later - still believed that he had only "popped out for a cheese and ham croissant" and would be back before long to put in a good afternoon's work.

I have always been rather partial to a civilised prandial anabasis myself, and was now doubly grateful for the enforced extended lunch-break caused by the pressure our group had put upon the culinary arrangements and the fact that our waitress had now disappeared across the road, had hitched her skirts up, tucking them into the leg of her knickers, and was now paddling around in the stream with a bevy of wildly excited and muddy small boys, looking for an elusive lamprey.

Eventually, I heard a torrent of Spanish yelled out of the kitchen window - presumably by the increasingly beleaguered cook - and saw our waitress scuttle back towards the restaurant so that she could resume her official duties.

An hour or so later the leader of the schoolchildren - a swarthy youth called Pablo, came back to me. He had a very sad look on his face as if some deep and unfathomable sorrow was haunting him. *"Senor, I am so sorry for the little eel".* I assumed that that meant they had caught one, but killed it by mistake. Never mind, a dead specimen was better than no specimen at all. *"Why are you sorry?"* I said.

"Because there is no little eel", he said, and not for the first time I realised that the language gap did not really convey the nuances in what one was trying to say.

In the meantime, it seemed that once lampreys had attached themselves to our lunchtime conversation, (presumably utilising their suckers) one could not prise them off, and the subject kept on worming its way into the conversation as if by magic.

We were talking about the brief, but fierce, tropical storm that we had found ourselves in the middle of a few hours earlier, and someone - I think it was "Cheese" - broached the subject of Hurricane Georges and the damage that it had done to the island. I looked out the window at the river where I had caught the lamprey, and saw that what had, only half an hour before, been crystal clear water was now a torrent of bilious looking mud. I expressed surprise that it had taken so long for the muddy water to have worked its way down from the ever more eroded mountainside above us, and was told that the water course was so twisty and turny that it took some time for the mud from the increasingly eroded hillside to reach the relative lowlands of Canóvanas. "Cheese" told me that in the aftermath of Hurricane Georges in 1998, there had been considerable land slips, and he was in the middle of extolling the increasing problem of soil erosion caused by illegal encroachment into the rainforest for farming land, when he stopped in mid-sentence.

Years ago, when my dog Toby was still alive, I always used to be amused when he was earnestly crossing the room about his doggy business, when he would suddenly stop for a scratch. He would then look around with a puzzled expression on his muzzle, having completely forgotten what he was doing in the first place. Until then, I had never seen a similar expression on a human being, but I could see that "Cheese's" train of thought had been interrupted in a similarly dramatic fashion.

"You know the caves at Aguas Buenas?" he asked me. I certainly did. These had been where I had first encountered cave snails, and I had been told by our guides on that occasion that deep within the cave system there was an underground river that was practically unexplored. I nodded assent.

"Cheese" told me that during the aftermath of Hurricane Georges an enormous amount of water had flushed through the cave systems and he had been told that *"large anguilla, and other strange fish"* had been flushed into the open-air river system. He had not seen them for himself, but had been told that they were pale in colour, and unlike anything that had been seen in the river before.

I was immediately interested. Cave systems across the world are the homes for a dazzling diversity of troglodytic lifeforms including blind salamanders, incredibly specialised sightless fish, and an array of invertebrates unlike anything that ever experiences sunlight. It looked as if something similar could be found in the subterranean river below Aguas Buenas.

Back in 1998, we had been told how dangerous and ill-explored these cave systems were, but now it seemed that there was a very real biological reason for us to plan a further foray there.

All good things have to come to an end, and after nearly two and a half hours, David realised that time was running away, and that we really needed to go back to work. So we paid the bill, got into our respective vehicles and went back up the hill in convoy.

Our destination was one of the most prosperous farms that I had visited on the island. I had been told that this was the location for one of the most recent chupacabra attacks, and that it was here, late in 1998, when a number of peacocks owned by a farmer had met their vampiric nemesis.

We drove up through the tall grassland, and once again I marvelled at this incredibly dense, and apparently impenetrable, habitat. *"You know what, Nick?"* I said. *"Anything could live in there"*.

"Happen it could", he grinned back at me, and I felt a warm glow at being one of the few people in the multiverse that could actually get *paid* for travelling to the ends of the earth with one of his best friends.

By the gate of the farm there was another small flock of the tiny Puerto Rican goats, and again I marvelled at the paradigm shift that these tiny creatures was forcing me to consider. We drove into the farmyard, but before we were able to disembark, David trotted over to us from his car, gave us a walkie-talkie, and told us to drive down to the bottom of a cart track situated on the far side of the yard, to turn round at the bottom of the hill and stay there awaiting further orders. At the bottom of the hill we did as we were bid, and looked out over the lush countryside. For all the world it looked like a verdant valley in North Devon, and only the aggressive little anoles defending their territory with xenophobic enthusiasm, and the boat-tailed grackles looking even more mentally deficient than usual, reminded us that we were many miles away from the rich farmland just inland from Bideford Bay. A cheerfully vacant looking youth wearing a *Slipknot* T-shirt and whistling the aforementioned band's loveable ditty "People = shit" came wandering past.

"Hola senors", he grinned at us, ambled across to a nine-bar gate, leant on it and started making some extraordinary noises. Well, they were extraordinary to Nick who was city born and bred, but I knew exactly what he was doing. *"What the fook's 'appenin' like? Is he some kind of wazzark? We don't have things like that in Pelsall you know...."*

I may have lived in a particularly unprepossessing suburb of Exeter for several years, but I am a country boy at heart, and I know a cowman when I see one, and this one was calling his cattle for milking in the same way as cowmen do across the globe. I suddenly felt very homesick, not for my little house in Exeter but for the village in North Devon where I grew up. A year or so later I told this story to my Dad, as he lay dying, and he told me that he had heard exactly the same noises made by Hausa tribesmen in Nigeria. There is something between a cowman and his herd which is universal; something primal which probably goes back to when the first Mesolithic hunter-gatherers decided that hunter-gathering was a bit too much fag, and that it might not be that bad an idea to try and domesticate one of them there aurochs they had seen on their travels.

Nick looked at me as if I was mad, and not for the first time, and certainly not for the last, I realised that there was a far greater gap between those members of the human race who choose to live alongside millions of their fellows and those who prefer a rural existence than there was between our new Death Metal-fan friend, and any of the farm workers I had known as a child in the wilds of North Devon.

Suddenly, between thirty and forty feral looking Friesian cows came galloping over the hill towards us. Nick looked somewhat perturbed at the thunder of hooves, but I was more confused by the other peculiar noise that I heard. I thought, at first, that our friend the cowman had a portable CD player on him and was playing one of the less tuneful ditties recorded by his heroes. However, I soon realised that the wall of feedback and incredibly distorted vocals were coming from the walkie-talkie on the dashboard in front of us. We must either have been in a radio blackspot, or our gallant director was summoning us with his head lodged firmly within a galvanised iron bucket.

Assuming, correctly, that we were being summoned, Nick turned on the ignition and with a procession of cattle behind us, we drove up the hill towards the rest of the crew. I haven't actually seen the finished product that was broadcast towards the end of 2004 on the Sci Fi Channel, but I can imagine that an establishing shot of us, and our bovine companions, appearing over the brow of the hill to meet our next interviewee, would have been quite an imposing bit of television.

The farmer was a genial looking guy in his mid-50s and he told us how, although a dairy farmer with the ubiquitous plantation of plantain trees on the side, he had kept peacocks for years purely for their ornamental value. The incident had taken place towards the end of 1998, and one morning, he had been checking his livestock, when he found four of his precious peacocks dead and apparently drained of blood. The only wounds that he could see were a triangular shaped hole in the skull of each of the animals. This was the first and only time he had encountered such a thing, and one got the impression that he - as a relatively affluent, and obviously educated man, had rather sneered at the stories he had heard from his poorer and less well-educated brethren. However, after these attacks he had taken an interest in the chupacabra phenomenon, and had become firm friends with Ismael Aguyao.

As if on cue, there was a cheerful shout from the other side of the farmyard behind us, and our friend Ismael strode over to join us. It seemed as good a time as any to interview Ismael about his involvement with the chupacabra experience, and despite being only sketchy familiar with each other's languages I think we got an impressive interview on film.

Unfortunately, Ismael was accompanied by his unpleasant friend - the man who had wanted enormous sums of money just to let us look at his photographs and video of the alleged chupacabra footprints. I did my best to ignore him as I finished interviewing the farmer and Ismael, but he glowered darkly at us from the shadows and kept on interjecting in muttered Spanish until one of the crew had to tell him to shut up. I asked the farmer whether he had ever seen a chupacabra. To my delight he gave me one of the best eyewitness descriptions that I ever managed to glean. He described a strangely rangy creature, some four foot high, with a crest of spikes on its head and a procession of spikes down its back. He said that it walked semi bipedally, but that on the few occasions when he had seen it, it had always been at night, and had been startled by his torch, and had scuttled on all fours into the darkness. Like so many of the more paranormally-minded witnesses, he described it as having glowing red eyes, but suggested that - unprompted - this could merely have been the effect of the large eyes of a nocturnal animal being illuminated by the beam of his torch.

Although, by this time I was becoming more and more convinced that we were dealing with a flesh and blood animal, rather than anything out of a post-Lovecraftian horror novel, I asked him whether he had ever seen any strange lights in the skies. He burst out laughing. *"Only the peasants and fools think that. El chupacabra.... he is just an animal. Just an animal, but we don't know what he is!"*

Off camera, Ismael was translating our conversation for the benefit of his surly friend. His friend was not at all impressed by the farmer's last comment.

"Hijo de puta!" he muttered, and spat expressively into a convenient cowpat.

The farmer winked at me. *"See what I mean?"* he said.

I asked the farmer how many times he had seen the creature and was gratified to hear that he had done so on four or five occasions. I was expecting to hear that the sightings had all occurred at about the time when the attacks had taken place, but was excited to find out that he had seen it on two occasions within the previous few weeks. On the most recent occasion he had been patrolling his plantain crop when he saw a chupacabra scuttling away from him into the darkness. The next morning he went to inspect his livestock. After a sleepless night, he was convinced that he would find that his peacocks would have been attacked again, or at the very least his goats or his chickens would have been decimated. Much to his surprise, he found nothing of the sort. His livestock were alive and well, but there did appear to have

The high grassland plateau where "anything could be living"

Farmyards are farmyards and cattle are cattle whether you happen to be in North Devon or Puerto Rico

The countryside was so familiar, it could have been my beloved rolling Devon hills rather than an upland farm in Puerto Rico

ABOVE: Interviewing Ismael at the farm BELOW: The paddock where the peacocks were killed

THE BANANA PLANTATION

On the insert picture of the tree that had been attacked one can see how the tree had been killed by the peeling back of the pithy layers of the trunk. You can also see how the undergrowth by the side of the tree had been trampled down by `something` presumably a medium sized animal.

豪猪 Hystrix hodgsoni
别名：中国豪猪、箭猪、刺猪、普通豪猪、刺猬

been one victim of his mysterious visitor. One of his prized plantain trees had been cut down. He took me to see it, and although it was rapidly rotting away under the influence of an excessively rainy Puerto Rican summer you could see that the peculiar pithy stem had been stripped away in much the same way as a squirrel will ring the bark of a branch of an English oak tree, condemning it to a slow and ignominious death. In this case, the death had been much quicker. Layer upon layer of the pithy stem had been chewed away until there was no longer enough substance to support the weight of the vast tree, which had collapsed in on itself. I had seen something like this before, many years ago, but did it have any bearing on the current case?

My father was really quite senior in the Hong Kong Government during the 1960s. He is dead now, and I believe that everyone else involved must be either dead or in their dotage by now, so although some of what follows should really have been allowed to remain buried, I don't suppose it really matters now, and it is an interesting tale with some quite significant bearing on the main thread of this narrative.

From 1841-1997 Hong Kong was a British Colonial Possession. On April 30, 1841, 12 weeks after the British had landed in Hong Kong, orders were given by Captain Charles Elliot to establish a police force in the new colony. The first chief of police was Captain William Caine, who also served as the Chief Magistrate.

As Wikipedia (the free encyclopaedia) recounts:

"For several decades Hong Kong was a 'rough-and-tumble' port with a 'wild west' attitude to law and order. Consequently many members of the force were equally rough individuals. As Hong Kong began to flourish and make its place in the world Britain began to take a dim view of the government's lack of grip in both public and private sectors, and officials with strong values and Victorian concepts of management and discipline were sent to raise standards. Strong leadership, both of HK and of the force began to pay dividends towards the latter part of the 19th century, and business prospered accordingly."

This conflict between frontier morals and those of the powers that be back in the motherland was a reocurrent theme in the history of the Royal Hong Kong Police. In the mid-1960s there was a massive corruption scandal within the force. Once again we should turn to Wikipedia:

"The spectre of corruption became prominent in HK in the 1960s, the HKP - as did almost every government department - experienced this and it peaked between 1962-74, involving officers of all ranks and ethnicities. Reasons? Motives and opportunities were many and and varied, but mainly - 'motives' (poor pay and worries about Red China invading and abolishing pensions), and 'opportunities' (HK was enjoying vibrant economic progress and its industrious, self-starter people were forming thousands of small street-level businesses all ripe for 'protection'.

During this time, the police, along with members of departments like Public Works, Fire, Transport et al all had their own distinct methods of earning illicit income. The police were the offenders with the highest profile and it took the determined stance of Governor MacLehose together with Commissioner Sutcliffe to instigate the firmest of measures to eradicate syndicated corruption - and the establishment of the Independent Commission Against Corruption (ICAC) in 1974 was the prime one. After teething troubles, including a mass walkout by officers in 1977, by the early 1980s a combination of the ICAC, firm police management, better emoluments and an amnesty had succeeded in destroying the overall culture, removing powerful figures, educating against greed and increasing accountability."

My father was one of the senior Civil Servants involved in leading the battle against the corrupt officers. I don't know the details, but I suspect that as my father was Assistant Financial Secretary, and Assistant Colonial Secretary at the time, that he was probably involved more in the process which eventually improved the pay and conditions of the force, rather than any undercover shenanigans involving lurking away in back alleyways with a revolver; but then again with my father, one could never be *quite* sure.

However, what I do know is that as a result he became close friends with both the Police Commissioner and his immediate deputies, and as a result my family hang out socially for a while with people who would normally have been several steps up the pecking-order from us.

People living in the England of the 21st Century find concepts like that both incongruous and disturbing. As we supposedly live in a classless society where everybody is equal, the idea that a society of Englishmen and women, well within living memory, could exist on such strict social rules. But it's true.

We don't live in a classless society now, despite the efforts of successive governments to try and tell us that we do. The only reason that the aforementioned succession of governments has done this is to prime successive generations of consumers into spending more and more money, to fuel the onward Gadarene progress of the juggernaut that is market forces. Ever since the liberal hire-purchase legislation of the early 1960s, British consumers have been encouraged to spend more and more money, get deeper into debt, and basically to live beyond their means, and the move towards so-called classlessness is merely a result of this. As a result the social order in Britain has broken down, and many of the current social disorders in the country as a whole are a direct result of this.

In the Colonial Administration that was the Crown Colony of Hong Kong during the 1960s, everyone - including the pre-teen me - knew where, and who, they were, and I truly believe that people were - on the whole - happier, for being part of a rigid social hierarchy.

As I have already written, the senior members of H.M. Government lived in luxury apartments on The Peak. However, the senior echelons of the Police and Prison Service lived in bungalows with gardens, and as a direct result of my father's involvement with the anti-corruption drive, for the first time in my life I found myself visiting homes with gardens. This was another great leap forward for my burgeoning career as an amateur naturalist.

On one unforgettable occasion we were having Sunday lunch with the family of one of the senior policemen, in his bungalow somewhere in the New Territories. I cannot remember - to be quite honest - whether it was the Police Commissioner himself, or one of his acolytes, and it doesn't matter really, but the important thing was, that like so many of his ilk both at home and abroad, he was a keen amateur horticulturalist.

This particular scion of the long arm of the law was particularly proud of his prize-winning banana trees. He was a Colonial Service Officer of the old school, rather like the one that my father told me about in Nigeria who had died of apoplexy after finding out that his houseboy had made his mustard with milk rather than brandy, and I remember his face florid with rage as he told my father that some *"bloody iconoclast"* had destroyed both of his precious trees.

He took us into the garden, where I still vividly remember seeing a pair of mutilated trees bearing exactly the same stigmata as did the one that I was to see in Puerto Rico thirty five years later. Three and a half decades apart, the two trees were identical.

The important thing is what happened next. It was 1968, and revolution was afoot across the globe. In 1968 the world turned upside down.

In Vietnam, US imperialism was being humiliated and millions of students and youth in the west were in open rebellion against the war and the system that generated it. As a child in Hong Kong I still remember the troopships full of demoralised American GIs straight from the Tet Offensive. They were supposed to be visiting Hong Kong for a few weeks' shore leave, and well-deserved (in their eyes at least) rest and recuperation. However, even at the age of eight I could see that whereas these poor battered wrecks of men had at least temporarily escaped the green hell of paddy field and burned out forest, their eyes betrayed the fact that they were in their own private hells – a place from which most of them would never escape.

From the pitched battle of London's Grosvenor Square, to the barricades in Paris and on to the riot outside the Chicago convention of the US Democratic Party, young people were openly challenging and fighting the establishment. 1968 brought an end to the political 'stability' of the post-war period and was an early indicator of the revolutionary events that were to follow in the 1970s. My family had gone on holiday to Scotland that year, and I still remember my father's abject horror when a random hippy in the street gave my little brother a copy of the 'Little Red Book' – the thoughts of Chairman Mao.

In Czechoslovakia, workers took to the streets in struggle against the tyranny of Stalinism, in Northern Ireland the civil rights movement mushroomed. In America itself, as well as huge anti-war protest, the black civil rights movement lost one of its leaders, Martin Luther King, at the hands of a gunman, but would develop to challenge the reactionary bigotry of the American political establishment.

After King's assassination many major US cities erupted. The biggest riot of them all was probably in Watts Town, where a sizeable area of Los Angeles was torched. In June, Bobby Kennedy, who may well have won the Democratic nomination for the presidential election on an anti-war ticket was gunned down at a meeting. And at the convention itself, in Chicago in August, the police went wild and attacked the anti-war demonstrators outside and inside the convention, with truncheons, maces and tear gas. The police dragged at least two delegates from the hall, and the riot reached a high point on the steps of the Hilton Hotel where the world's television flashed the pictures of the brutal beatings being meted out by the Chicago police force.

In Britain it was a year of student protest and occupations. A year of cultural turmoil and experiment, with bands like the *Beatles* at their greatest and most powerful. It was also a year of the first political stirrings against the right wing policies of the Labour government.

But the revolution had started in China two years earlier. In China, The Great Proletarian Cultural Revolution in the People's Republic of China was a struggle for power within the Communist Party of China that manifested into wide-scale social, political, and economic chaos, which grew to include large sections of Chinese society and eventually brought the entire country to the brink of civil war. It was launched by the Communist Party of China's Chairman, Mao Zedong on May 16, 1966, officially as a campaign to rid China of its "liberal bourgeoisie" elements and to continue revolutionary class struggle. It is widely recognised, however, as a method to regain control of the party after the disastrous Great Leap Forward led to a significant loss of Mao's power to rivals Liu Shaoqi and Deng Xiaoping, and would eventually manifest into waves of power struggles between rival factions both nationally and locally. Between 1966 and 1968, Mao's principal lieutenants youth militia called the Red Guards to overthrow Mao's perceived enemies and seize control of the state and party apparatus, replacing the Central Committee with the Cultural Revolution Committee, and local governments with revolutionary committees. In the chaos and violence that ensued, many revolutionary elders, authors, artists, and religious figures were purged and killed, millions of people were persecuted, and as many as half a million people died.

It is a little-known fact that there were Red Guards in Hong Kong. There was a revolution in Hong Kong in May 1967, and it was a revolution that very nearly succeeded! What are now known as the Hong Kong 1967 Riots, but what was in fact an attempt by the Peoples Republic of China to overthrow the Hong Kong government, started in May 1967 when local communists turned a labour dispute in an artificial flower factory, into large scale demonstrations against the rule of the Gwai-Lo.

On 8th July, Red Guards killed five policemen in a gun battle at the border town of Sha Tau Kok. Many years later my father told me that - together with the other senior members of HM Government, he had been called to a secure bunker below the Government Offices. There, he and his colleagues had been issued with side arms and told to expect an imminent invasion.

This was not as much of a shock to me as you might imagine. One of the most horrific moments of my childhood happened when I wandered into my parents' bedroom one night. I am very aware that this sounds terribly familiar to anybody who has read an account of someone talking about *their* traumatic childhood. It is usually followed by an account of how the unfortunate child had witnessed either his or her parents indulging in sexual acts which imprinted themselves on to the darker parts of his or her emerging psyche, or - sadly - when the poor child witnesses some horrific act of domestic violence. In my case it was much worse. Although I never told either of my parents this, I both saw, and heard, my father give my mother a revolver with three bullets; one for my little brother, one for her and one for me. Stuff like that really *does* have a negative effect upon a nascent psyche.

In the event, the invasion didn't happen. There were lots of bomb threats. Real bombs, and even more decoys were planted throughout the colony. Laboratories in some leftist schools were turned into bomb making workshops.

In response the police fought back and raided leftist strongholds. In one of the raids, helicopters from *HMS Hermes* were called in to land police on the roof of a 20-plus-storey building. Upon entering the building the police discovered bombs and weapons as well as a 'hospital' complete with dispensary and an operating theatre.

A seven-year-old girl and her two-year-old brother were killed by a bomb that had been wrapped like a gift and placed on their doormat, and a bomb was planted - and disarmed by the army - in the playground of Mount Austen Mansions, where we lived at the time. I remember being surprised by how quiet the ensuing explosion was.

I only saw one troop of Red Guards. They were marching up the Peak Road outside our flat and chanting something about Ho Chi Minh.

Against such a background, is it any wonder that my father's friend, the senior policeman, was somewhat paranoid and - as it turned out - overreacted to an almost comical degree? He had managed to convince himself that the damage to his precious banana trees had been an anti-imperialist gesture by local communists, and so, in what might these days be seen as a flagrant abuse of his position, but which even now I think in the light of what had so nearly happened the previous year, was quite probably totally justified, he positioned two Special Branch marksmen with high powered rifles to guard his garden in case of further attacks on his property.

Hong Kong, like Britain, did not have a history of armed policemen, and at such a tender age I was terribly excited to see real *guards* with real *weapons* guarding my father's friend's house. We went home, but - so I discovered later - that very night the forces of misrule once again entered the senior policeman's garden. This time, they were no match for the forces of the British Empire, and they were summarily shot. There were two of them. They were not communists (for rodents, no matter how large, are not known for their revolutionary politics). They were porcupines.

I quote from the official Hong Kong Biodiversity website:

"**East Asian Porcupine *(Hystrix hodgsoni)*** is a large heavily built rodent, with a short tail and with some rattle quills. The front of the body has short, dark brown spines, while the hindquarters have hollow, long, pointed quills which are conspicuously marked with black and white bands. There is a short whitish crest on the neck and upper back. The tail is short with both long, pointed quills and rattle quills. The species is common and widespread throughout Hong Kong, but no record has been found on Lantau Island".

The whole affair was a five-minute wonder in Hong Kong society. Although my father's friend was quite understandably embarrassed by it, nobody really blamed him. He was later honoured by the Queen for what he had done to prevent a communist takeover (not for services to porcupines) and indeed in 1969, Queen Elizabeth granted the Hong Kong police the privilege of being the Royal Hong Kong police – a title that they kept until 1997 when the colony reverted to Chinese rule. But most of this passed me by at the time. I was more upset at the deaths of the two porcupines, and thirty-five years later at the tiny farm high in the mountains of Puerto Rico, the whole affair came rushing back into my head as I realised - once and for all - the identity of the strange animal that had attacked my new friend's plantains.

There was, however, one big problem. There are no porcupines known in Puerto Rico.

The problem was actually far more impressive than that. There are twenty-seven different species of porcupine in the world, but they are divided into two distinct families. The eleven Old World porcupines in the family, Hystricidae, are almost exclusively terrestrial, are relatively large, and - like the animal described by the Puerto Rican farmer - sport clusters of relatively large quills.

Thirty million years ago, they separated away from the New World porcupines, or Erethizontidae, which are much smaller, arboreal, and have their quills attached singly. They are far less nocturnal than Old World species, and have long, powerful, prehensile tails which help them balance when they are feeding off bark, leaves and conifer needles high in the treetops. There are no known porcupines anywhere in the New World that resemble any of the Hystricidae. [1]

Although my brain was working overtime at this point, I decided that it would be unwise to share my revelations with any of the locals or anyone from the crew. So - making a mental note to confide in Nick as soon as I could - I changed the subject, and asked the farmer about his dead peacocks, and - using my walking stick - he drew a diagram in the dirt showing the triangular hole in the skull through which the brains had been eaten. Once again, this rang alarm bells in my head.

But this was neither the time nor the place to explore such thoughts, so I placed them firmly into my mental filing system for later use.

When we had concluded the interview we drove back to Canóvanas. We gave Ismael and his unpleasant friend a lift in the back of the Jeep. The sky looked bright and blue, so we didn't bother to put the covers up and drove back across the grass filled plateau until, as dusk was beginning to fall, we could see the twinkling lights of Canóvanas below us in the valley. It began to rain and by the time we pulled in to the car park of the bar where we had met Ismael previously, all four of us were soaked through.

Somehow, by dint of the sort of personal voodoo that camera crews across the world seem to have, the crew were before us. Such was the level of camaraderie that had developed over the previous few days

---

1. The material on the history and prehistory of the porcupine family comes partly from Wikipedia, partly from various animal encyclopaedias and mostly from long conversations with the late Clinton Keeling, Chris Moiser and Dr. Darren Naish. Mostly from Doctor Daz.

that our entrance - soaking wet - was greeted with ribald comments and cat calls.

We felt that we had been given the status of regulars at this tiny bar, and - as it was our last day's filming on the island - we sat down at one of the formica topped tables and bought a round of drinks for all the locals who had been so kind to us.

Suddenly Redfern began to snigger. We both have very childish senses of humour, but I could tell from the tone of his guffaws that something particularly juvenile had taken his fancy. I turned round to him and could see him shaking with laughter, as he pointed at an album cover that had been pinned above the jukebox. It was an LP by an artist called Willy Colon. A name like that was irresistible and I, too, burst into helpless laughter.

From Wikipedia (the free encyclopaedia):

William Anthony Colón (born 28 April 1950) is a Puerto Rican salsa music icon. First and foremost a trombonist, Colón also sings, writes, produces and acts. He is also involved in municipal politics in New York City.

Willie Colón was born in the Bronx, New York, to Puerto Rican parents. He picked up the trumpet from a young age, and later switched to trombone, inspired by the all-trombone sound of Mon Rivera and - at least during a specific period in the 1960s - that of Eddie Palmieri. He was bullied in his neighborhood because of this, and had to defend himself quite often from gang members. In a way similar to Bob Marley's, he gained a reputation for being tough and fierce in combat, even if his height could put him in a disadvantage (Willie is 5ft. 6in. -168 cm- tall). He spent some summers at his maternal grandmother's farm in Manati, Puerto Rico, where he claims he learned the discipline and tenacity to thrive on his own, as well as a strong love for Puerto Rico that shaped his later political views (he's a strong supporter of Puerto Rican independence and Latin American political unity, and theoretically a liberal, even if later dissapointments with the Democratic Party (United States) of the United States moved him to endorse Republican candidates.

He was signed to Fania Records at 15 and recorded his first album at age 17, which ultimately sold more than 300,000 copies. Due to fortuituous events (he had no singer for his band at recording time since his first candidate was killed in a street fight), the main record producer at Fania at the time, Johnny Pacheco, recommended Héctor Lavoe to him. This led to a very successful collaboration between the two, which ended in 1975 when Willie quit touring to raise a family and pursue various business ventures (including computer programming) to guarantee a steady income at home. Willie did produce some of Lavoe's solo records afterwards.

Beyond his skills on the trombone, he rapidly excelled as a composer, arranger, and singer, and eventually as a producer and director. Combining elements of jazz, rock, and salsa, his diverse work incorporates the rhythms of traditional Cuban, Puerto Rican, and Brazilian music.

He went on to have many successful collaborations with salsa musicians and singers such as Ismael Miranda, Celia Cruz and Soledad Bravo, and singer-songwriter Rubén Blades.

In September 2004, Colón received the Lifetime Achievement Grammy Award from the Latin Academy of Recording Arts and Sciences, but the fact is that he has a name that *Viz* magazine's Finbarr Saunders would have found highly amusing!

Laughter is infectious, and although none of the people in the crew - or the burgeoning crowd of locals who had gathered in the tropical twilight to share a beer with the eccentric, but obviously well-meaning *locos gringos* - understood what we were laughing at, they all joined in, and soon we were in the midst of a very convivial, and very good-natured, party.

But we weren't just there to have fun; we were on the track of a monster, and so before long, we got back

Filming in the cafe

ABOVE: Devin Macanally shows the author the skull of the blue dog (close-up below)

ABOVE: Devin shows the author around his ranch
BELOW: The author examines the shoulder blade of the unfortunate beast

ABOVE: The Elmendorf beast when alive (courtesy Whitley Strieber) BELOW: A dog suffering from the pattern of sarcoptic mange typical from the San Antonio region of Texas. Note how the pattern of hair loss (if, indeed, the baldness of the above animal *is* hair loss) is completely different. Having taken veterinary advice, it would appear that a dog that was so riddled with mange as to cause this amount of hair loss would appear otherwise far more unwell than this animal did; it was apparently completely healthy until Devin shot it.

to business. Talking to Ismael, I discovered something very important, and something that - to the best of my knowledge - has never been documented before.

Many authorities, including me, have noted that the name chupacabra is not just used for the infamous goatsucker of the Canóvanas plateau. It is an excessively offensive term for the most degenerate prostitutes (one who would degrade herself so low as to, literally, 'suck' a goat) but that as a term - indeed one of the vilest terms - of abuse, it has in the past, been used to describe fearsome monsters. However, the use of the term to describe the spiky creature of the Canóvanas plateau goes no further back than 1995 when it was coined by none other than Ismael and his unloveable friend. This piece of information does much to negate the currently popular theory that the chupacabra is some kind of ancient daemon that has been known across Latin America ever since the arrival of the Europeans, and probably before. That, I am afraid to say, is pure nonsense.

Six months later, I found myself in San Antonio, Texas, as a guest of the *Discovery Channel*. I was filming the pilot episode of a series, which - in the end - was never made.

In the early spring of 2004, Devin Macanally, a rancher who owned a smallholding near Elmendorf - a small town outside San Antonio in Texas - saw a strange bluey-grey dog-like animal scavenging for mulberries that had fallen from a tree in his garden. Macanally - an amateur naturalist - watched the creature for a while, and then forgot about it. He saw it again several times over the next months, and when thirty-five of his domestic poultry, and at least one of his heifer calves were killed by an unknown predator, he did what so many other farmers have done across the globe when confronted by an unknown animal that they believed had been killing their livestock, and the next time he saw it, he reached for his rifle and shot it.

A picture of the strange, wizened blue-grey corpse was flashed across the wire services and soon ended up in the cryptozoological newsgroups on the Internet. Nobody knows who it was who first suggested that the animal was a chupacabra - it certainly wasn't Macanally, because in one of the earliest news stories about the incident he is quoted as saying:

"First thing that came to my mind, is surely everybody's gonna think this is a chupacabra. But it's so odd because it has no hair."

He told reporters from the beginning, as he told me the following November, that he had no doubt that this was some kind of strange dog-like creature, and that there was nothing the slightest bit paranormal about it. However, the damage was done and another news story quoted:

"At the nearby Deleon's Grocery and Market, customers come in to check out pictures of it. One woman says it is exactly how her grandmother described the dreaded chupacabra."

Well, we know that this is nonsense, if only because we have conclusively proved that the term *chupacabra* was not even coined until a decade ago by my mate Ismael back in Canóvanas . However, the story spread across the internet like wildfire, and as always, everyone with a vested interest got on board and decided to voice their opinions about it. Whitley Strieber - a gentleman whose main claim to fame is that he made a fortune telling the world about his nocturnal anal adventures with aliens, contacted Macanally, and took away bones and/or tissue for DNA analysis. Macanally heard nothing for some while, but told me that he was "disappointed" when the first that he heard of the test 'results' was when Strieber announced that the DNA "had deteriorated due to exposure to light, heat or radiation" and was unidentifiable.

This announcement provoked a predictable backlash, with every kook online immediately deciding that this was proof either that the Elmendorf beast had been the result of some horrific government experiment, or was some 'entity' from another galaxy or dimension. Luckily more sober viewpoints prevailed, and most people in the cryptozoological community were prepared to dismiss the creature as nothing more than a coyote with mange! This was - I have to admit - the view that I was quite prepared to take, until November 2004 when I received a telephone call from a TV Producer in Los Angeles who was interested in having me present a series on paranormal investigation for the *Discovery Channel*.

Now, I ain't blowing my own trumpet, but as one of the few people in the world who actually ekes out a semblance of a living as a fortean and cryptozoologist, this sort of thing happens all the time. Invariably - and this turned out to be no exception to the rule - they fly me out to make a pilot episode, find out that I am not a wild eyed member of the "I want to believe" brigade, and I never hear anything from them again. So, I have become used to doing my best to get what I can out of the initial encounter, knowing full well that I am unlikely to get any proper paid work out of it. On this occasion I managed to inveigle the production company into flying my then girlfriend and me out to Texas for five days to shoot the pilot (and have a free holiday).

It was only when - after a series of ludicrous misadventures - we finally arrived at San Antonio airport eight hours late that I discovered that the subject of the pilot was to be the Elmendorf beast. *"But its only a bloody coyote with mange"* I protested.

"No its not. It's the chupacabra", said the producer, and glared at me.

"Jesus H Christ" I muttered, and my then girlfriend who was a staunch Mormon glared at me.

Two days later, we found ourselves driving the producer's shiny new 4x4 up the battered cart track that led to the Macanally Ranch. Having come from a generation who were brought up on the TV westerns of the 1960s I had always imagined 'ranches' to be rolling acres of flat prairie populated by the occasional buffalo and big-breasted women waving six-shooters. During my various trips to The States I have visited numerous ranches in several states and to date none of them have even slightly resembled the High Chapparal. This one was no exception. Macanally - who was a pleasant looking man of slightly wizened middle age - lived in a single story wooden building surrounded by overgrown bushes, and with a number of large (and slightly dilapidated) aviaries containing various motley domestic fowl, and a couple of magnificent peacocks.

Macanally was a little reticent at first. He explained that he had become sick to death of being plagued by the lunatic fringe who kept on coming up with more and more ludicrous explanations for the beast, and felt that the tone of some of the news stories that had appeared about the creature had not shown him up in the best of lights. Later - after I had gained his confidence, to a degree at least - he explained that being of mixed Scottish and Irish ancestry he had little love for the English, and had initially thought that I was just some crazy longhaired limey out to make a fast buck out of his story.

I managed to convince him of my credentials as a cryptozoologist, and he showed me the location where he had first seen the animal. He described how on the occasions that he had seen it, how it had moved in a completely un-coyotelike manner. *"You can always tell coyotes by the way they walk. There's hundreds of them on the ranch here and I see them all the time. This animal was different. Not only was it bald and blue, but it moved in a completely different way. It was more like one of the jackals that you see on the Discovery Channel"* he said. He also told me - much to my surprise - that his friends and neighbours had seen similar animals for some years.. Unlike coyotes, they hunted by day, and usually hunted in pairs, and the pattern of hair loss - if indeed it *was* hair loss, and not a hair pattern specific to this unusual creature - was identical on each of the creatures.

This put a completely new slant on things. What made it even more interesting was the story that I heard from a number of witnesses how on at least one occasion a mother had been seen with two cubs - all three animals being bald and blue. And according to Devin Macanally, the animal he had shot had been in the second trimester of pregnancy at the time.

I asked him what had happened to the body. I knew that it had been buried, and was prepared for the unpleasant job of exhuming the rotten carcass. Much to my pleasure he told me to wait there for a moment, and disappeared inside his house, to emerge a few minutes later with a large black bin-bag full of bones, which he emptied - in a cavalier fashion - onto some old newspapers in the back of his pickup truck.

Handling the bones I could see that this was no deformed creature; they were perfect, but there was something unsettling about them They seemed harder and more resilient than most canid bones that I have handled, and the skull seemed to have what appeared to be a small sagittal crest on it.

He showed me other skulls of coyotes that had been shot on the farm. One was much as you would expect, but the other - far larger than the skull of the animal that he had shot earlier that year - also had a pronounced sagittal crest.

This is where I finally cooked my goose with the film crew. I did - what I have to admit was one of my best performances ever - a deadpan piece to camera. *"We came here looking for the corpse of the chupacabra. Not surprisingly we found no such thing. What we have here is something far more exciting. There is a great misconception in cryptozoology that all unknown animal species are prehistoric survivors. What people forget is that evolution is a continuum. What we have here MIGHT be an example of evolution in action. We might be seeing the beginnings of a new species evolving away from the coyote species as a result of the environmental pressures of life at the beginning of the 21st Century".*

The producer looked at me.

"So, its not anything paranormal?"

"No, of course not", I laughed, and I could see my burgeoning TV career disappear in smoke.

I have no idea what the Elmendorf beast actually is. I am fairly convinced that it is *not* a coyote with mange. Whilst in the area I took a series of photographs of local dogs suffering from both main sorts of mange - Demodectic mange is caused by *Demodex canis*, a tiny mite that cannot be seen without the aid of a microscope, and Sarcoptic mange, also known as scabies. I was surprised to see that the pattern of hair loss looks nothing like those noted from the Elmendorf creature.

Although I have my doubts about the veracity of some of the other reports of hairless dogs from Texas - the one from Lufkin was obviously seriously ill and *did* appear to be suffering from mange, I feel that, especially in view of the accounts of these creatures seen alive across the region, that it would be unwise to dismiss them all as mangy coyotes. Whether or not we are indeed seeing the beginning of speciation in the coyote tribe, something strange is happening in *'them thar hills'*, and Devin Macanally has invited the CFZ to come back and investigate further. An invitation we are determined to accept.
But that is another story.

I visited Deleon's grocery store, and the woman there, Dolores Sanchez (although I have a sneaking suspicion that her giving me that name was somewhat akin to an English witness who wanted to preserve

The skull had a pronounced sagittal crest

her anonymity calling herself 'Sue Smith') repeated her claims that the blue dog was the "chupacabra" described to her by her maternal grandmother. Everyone in the area had known about this beast for generations, she told me, and they were all terrified of it. Yes, indeed she had heard about the chupacabra of Puerto Rico, and was convinced it was the same thing. I bit back the temptation to tell her that she was talking nonsense, and came - not for the first time in this book - to the conclusion that, as Lazarus Long was once reported to have said, one should never underestimate the power of human stupidity.

Back in the Canóvanas bar, Ismael's revelation proved once and for all that such claims were nonsense.

The word must have spread across the small town that there were scientists from Europe who were prepared to take the chupacabra account seriously, and within half an hour we were surrounded by a string of potential witnesses.

One very strange looking man - who for some reason reminded me irresistibly of a Latino analogue of Captain Mainwaring of *Dad's Army*, as portrayed by the immortal Arthur Lowe, told me how one night, in 1997, he had been driving his truck along a deserted country road, when he heard an unearthly sound. Not at all to my surprise, he described it as being like an enormously amplified heartbeat. He told me how the normal sounds vanished and how all he could hear was this relentless, monotonous, drumming. Suddenly, a black creature that he described as being 'flat', bounded into the road before him like *"'ow you say? A wallaby-a-roo ... like, you know, Skippy on television"*. He said that he saw this peculiar creature for several seconds before it bounded into the air, unfurled bat-like wings, and flew away.

Had it not been for his description of the "heartbeat", I would have been tempted to file this story away, alongside the accounts of the five-legged creature from Aguas Beunas about which I had been told on my previous visit to the island, and which sounded for all the world, like the creatures described in *The Mote in God's Eye*, a science fiction novel by Larry Niven and Jerry Pournelle, which was first published in 1974. He discovered creatures nicknamed 'Moties', which - like the animal described to me back in 1998 - was completely alien to any notions of tetrapod zoology.

However, I too, had experienced a sighting of something that could not be explained using orthodox terms of zoological reference, and I, too, had heard the relentless thrumming of what could only be described as an amplified heartbeat.

I offer no apologies for quoting here from my 2004 autobiography *Monster Hunter* in which I describe the events of January 2003, at Bolam Lake in Northumberland:

"As the dusk gathered at about 5 o'clock, I again heard the raucous noise of the rooks that he had reported, just before dawn. Suddenly, once again, they fell silent and one of the *Twilight Worlds* members shouted that she could hear something large moving around amongst the undergrowth. I ordered all of the car drivers present to switch on their headlights and to put them on to full beam. He did not hear any noise in the undergrowth although other people present did. Eight people were watching the woods and five of them, including Jon saw an enormous man shaped object run from right to left, disappear, and then a few moments later run back again.

It was an incredibly weird experience. I have been playing it back in my mind over and over ever since. It wasn't the first time that I had seen a cryptozoological creature. I had, after all, seen the Beast of Bodmin back in 1997, another mystery cat 10 years before, and I had even held the Mystery Lamprey of Puerto Rico in my hands. However, although I had encountered things which I had been unable to explain using strictly zoological and scientific terms of reference at second hand before, for example the animals that had been killed by the chupacabra, this was the first time I'd encountered a monster for real.

What I saw was an incredibly flat and angular figure, which appeared to be two-dimensional. It is here that I have greatest difficulty in describing my encounter because what I saw were so far out of the normal run of human experiences that there simply are not adequate words to describe it. I will do my best, but even now - over a year after the expedition was done and dusted - I still find it almost impossible.

As far as its shape is concerned, the nearest analogue that I have been able to come up with is the angular metallic running man which can be found in certain levels of the computer game *Doom II*. However, this "thing" (as Ivan T Sanderson would undoubtedly have called it), was a matt-black. It was so black in fact that it was a quasi-human piece of nothingness, which had somehow become projected upon the Northumbrian landscape. It moved far too sharply and far too fast to be a living thing - at least in the ways that we know it. Again I have to resort to another unwieldy and fairly inadequate analogue. It was as if somebody had filmed this humanoid shaped piece of nothingness, and then projected it back on to the landscape using the fast-forward facility on the video recorder. The whole experience lasted only a second, but it has been with me ever since".

I needed no convincing that weird shit can - and indeed does - happen. However, I remained convinced that the undoubtedly paranormal incidences and the sightings, of what I was getting more and more convinced, were a *bona fide* unknown animal, had very little to do with each other.

The next witness was refreshingly down to earth. He was a tiny, 'little old man', who looked as if he was about 100, although having seen the way that the ageing process is particularly unkind to those people who live near the breadline in underdeveloped countries, he could have been anything over 60. He hobbled over to meet us, as if every step was an effort of Herculean proportions, and he wheezed as if every breath could be his last. But when he opened his mouth to speak, I was amazed to find that he had a voice like the actor Brian Blessed (with whom I had a memorable drinking binge following an appearance on the Ester Rantzen Show in 1999). He boomed, and carefully enunciated every word, even though he was speaking in a peculiar patois somewhere between Spanish and pigin-English. Many of the words he used were completely unfamiliar to me, and it is fair to say that if I had not had Ernesto as my interpreter I would not have understood one word in twenty. I found out afterwards that he came from a very old family, and that although he would probably have denied it, it was popularly believed that he had far more Taino Indian blood coursing through his veins that he would care to admit. Even now, it appears, that a native Caribbean heritage is not one of which people are overly proud. He told me how – nearly ten years before – his flock of chickens had been amongst the first to have been visited by the chupacabra. He told me how, in the middle of the night, he had been roused by a commotion in his chicken run, and had gone out to investigate. He did not see the predator, but - or so he said - the reality of what he had seen that night had haunted him ever since. Eight of his chickens, lay lifeless in their coop. Although the wire was intact, it had been squeezed so there was a gap little smaller than his fist in the side. He was sure that the gap was far too small for a four foot creature like those described by his friends and colleagues to get through, but nevertheless, *something* had killed his fowl by biting a triangular hole in the roof of the skull, and eating their brains.

His description was so vivid that the niggling doubts I had experienced earlier began to feel stronger and stronger.

Then, finally, there was just Nick, me, Ismael, and his unpleasant friend left. The surly looking bloke grimaced at me in a threatening manner, spat vociferously into a large brass spittoon by the wall, and dropped the price that he would charge for us to use his alleged pictures to $50. Finally - much against our better judgement - we agreed to go to his house and examine them.

Alarmingly, Ismael had to go, and so we said our goodbyes, embraced and vowed eternal friendship, and followed the surly-looking man into the tropical night.

It turned out that he lived in one of the most notorious districts of the town - *Villa sin Miedo* (or Village

Without Fear).

The area is owned by the Puerto Rican government and designated as a relocation area for families of lower economic income. As in many impoverished areas, drugs are fairly common.

In 1983, the area was the scene of an infamous incident. During a police intervention, one policeman was shot to death, with a bullet to his lung. For a long while after that, the name of Villa sin Miedo was a household name all over Puerto Rico. According to area neighbours, the police used interventions like this to burn houses and destroy property that belonged to residents.

There have been several attempts to improve the quality of life of Villa sin Miedos residents. Many Puerto Rican, American and other international organisations have tried to help, with projects and cultural activities. One Brazilian professor who visited in 1993 noted that, during his trips to 80 or more countries, he had never seen people with so much hope as the people of Villa sin Miedo.

Driving to Villa sin Miedo was a particularly disquieting affair. Although I had found the streets of old San Juan in the immediate vicinity of the Windchimes Hotel to be vaguely unsettling, this place was excruciatingly unpleasant. I have been in unpleasant places before. I have been to Blackpool, in Lancashire, I have been to a village in Mexico whose squalor defied description, but driving through the tiny back streets of Villa sin Miedo was one of the scariest experiences of my life.

In Exeter, there is a district called Burnt House Lane, which is widely accepted as the roughest and nastiest part of the city. It is a place where the criminal underclass rule, where police never venture alone, and hardly ever get out of their cars, and where - whenever you venture there - you keep your hand on your wallet and one eye over your shoulder. If you can imagine a cross between this, one of the nastier sink estates in post-Thatcher Britain, crossed with one of the back streets of Venice where - when I visited it with my parents in 1971 - I remember my mother saying that she felt underdressed because she didn't have a cloak or a dagger, add the stench of an open sewer to the already foetid smells of a tropical night, and multiply your resultant paranoia a thousand-fold, and you might have some inkling of quite how unpleasant this particular part of Villa sin Miedo actually is.

We pulled up outside a surprisingly neat and tidy bungalow, and our companion disappeared inside.

About five minutes later he came out, clutching a handful of battered looking Polaroids. He showed them to us, and pointed out what he said were the reptilian three toed tracks of the goatsucker. Although we examined them by torchlight, we could see no such thing, and told him so. What about the videotape, we asked him? But, much to our horror, he turned on us like an angry rat, and started to shout. As I have explained elsewhere, my Spanish is rudimentary at best, but from what I could understand he was claiming that we were foreign filth, there to exploit him and his countrymen, and that if we did not go immediately, he would kill us.

"Fuck this for a game of soldiers", I said to Nick, and we left.

Driving through the almost deserted streets, we saw rats every few yards; enormous things like bloated maggots, and, as we turned a corner, I saw the unmistakeable silhouette of an old "friend" of my childhood. Another part of the puzzle was beginning to fall into place.

We were feeling surprisingly melancholy as we drove back down the now familiar freeway from Canóvanas to San Juan for the last time. This was our last night in the Island of Paradise, and although it wasn't as if wither of us were going back to ordinary suburban existences, our adventure - or at least this leg of it - was nearly over.

Island of Paradise

The tropical night is almost indescribable using terms of reference that someone who has never experienced one can understand. When night falls in temperate countries, even the most banal of landscapes can become beautiful and mysterious, enshrined in a shroud of darkness. But the tropical night is something completely magnificent. The darkness has the rich, warm, patina of a velvety plum.

Driving through the gathering tropical night is a truly magickal experience, and one that was made even more poignant by the knowledge that in about 28 hours time I would be back at home in Exeter. I always feel oddly emotional at times like this, and my reverie could not even be dented by the sound of Redders singing loudly along to *The Angelic Upstarts* as he drove us home at breakneck speed.

We got back to our hotel to find that most of the crew was sitting in the bar waiting for us. It was a strange feeling; we had only known each other for a few days, but there was an undeniable feeling of melancholy as we realised that this was - after all - our last night together.

"Cheese", sitting in the corner of the bar, was grinning even more than usual, and as I went over to sit with him, I realised quite how much I was going to miss him and the other guys when I returned to England.

He had brought me a present; a CD of jazz music from the island. The previous day whilst we had been sitting at the tiny Canóvanas restaurant waiting for the local children to come back with news of strange eels, the two of us had got into a deep conversation about Puerto Rico and its cultural history, and "Cheese" had promised to let me have some of the island's music. Thanking him profusely, I bought a round of Medella, and continued my conversation with him about the history and culture of the Island of Paradise.

Something of which I really had not been aware, were the cultural links between Puerto Rico and Cuba. If I had thought about it, which I must admit I haven't, I would have assumed that because Puerto Rico was a Commonwealth of the United States of America, that the islanders would share the distrust, indeed the paranoia, which the mainland USA has held towards the only Marxist state in the Caribbean since the adventures of the Bay of Pigs four and a bit decades ago.

However, as I was to learn, the cultural links between Puerto Rico and Cuba go far deeper, and since the days when they were the two jewels in the crown of the crumbling Spanish Empire, there has been an inexorable and difficult to define link between the two islands.

Cuba has been independent from Spain since the end of the Spanish American War in 1898, but the complicated history of the Cuban struggle for independence is a difficult one to relate. It was normally independent from the end of the war, but the United States imposed a 20-year treaty upon the fledgling republic, which in essence meant that the island was under United States control.

When the new president, Theodore Roosevelt, came to power in 1901 the situation changed again. Roosevelt had been a soldier during the Spanish American War, and was broadly sympathetic towards the viewpoint of those who wanted Cuban independence. He abandoned the iniquitous 20 year treaty, and gave the country almost full independence, with the leader of what had been the Cuban independence movement installed as President. However, even then, the new Cuban constitution gave the United States rights to intervene in Cuban affairs, and supervise its finances and foreign relations.

For the first half of the 20th Century Cuba limped on under a pro United States administration, and it was only following the Communist revolution, and the inauguration of President Fidel Castro, that Cuba began to break away from the United States. However, even then, there was an unwelcome American presence on the island. Following the terms of the Cuban-American Treaty of 1903, the Americans leased the area around Guantamamo Bay in perpetuity as a naval base. This has been a thorn in the side of the Cuban administration ever since, but they have been unable to do anything about it. Fidel Castro has refused to cash in any of the cheques which are the American government have sent for the leasehold of the property, apart from the first one in 1959. However, the Americans have retained possession of Guantamamo Bay ever since.

Following the Communist takeover of the island, there has been a trading ban in force, and - even though in the rest of the world, Communist and former communist countries have seen a distinct thaw in their relationship with the United States - relations with Cuba remain frosty.

On Puerto Rico, however, events took a different course.

Puerto Ricans have been U.S. citizens since 1917 by a statutory law called the Jones-Shafroth Act (or Jones Act) which can be modified at any time by the U.S. Congress (however, the citizenship of current U.S. citizens could not be revoked, only the status of those born in Puerto Rico in the future). They are free to live anywhere within the U.S. without a visa. Similarly, all U.S. citizens have the right to migrate to Puerto Rico without a visa. Puerto Ricans have no voting representation in the U.S. Congress and do not vote for the U.S. president. They don't pay federal income tax, but pay Social Security taxes. Puerto Rico has an elected Resident Commissioner, who sits in Congress as a delegate of the people of Puerto Rico - the delegate may speak in the United States House of Representatives and serve in committees, but may not vote.

"Cheese" told me about Puerto Rican politics. There are three main parties; each aligned to a different vision of the future of the island. One party desires full independence, another wishes to be granted full statehood within the United States, and the other wants to preserve the status quo with a few added freedoms.

When asked to choose between independence, statehood, or continuation of the present status with enhanced powers, as proposed by the Popular Democratic Party, Puerto Ricans have voted to remain a commonwealth. However, dissatisfaction with the current status is evident. The issue is still being debated and is on the agenda of all the political parties and civil society groups. Many pro-commonwealth leaders within the PPD are proposing an Associated Republic or Free Association similar to that in the Marshall Islands or Palau. The left wing of the PPD has achieved some success in driving the party to a less conservative and more nationalistic stance.

However, he told me that Puerto Rico feels that its strongest links are with Cuba. The clues are everywhere on the island - you only have to look at their two flags; they are of practically identical design. At this point another one of the crew, the swarthy young man called Pablo joined us, and told me that in his opinion the political future for the island, was to join in some degree of loose Confederacy with Cuba and possibly also with the two republics on the island of Hispaniola. He went on to explain why, and this was the thing struck me most profoundly.

As I have written elsewhere in this book, I am a child of the last generation brought up under the British Empire. I have always been broadly sympathetic towards the concept of the British Empire, and have not really understood it when successive generations of media pundits have described it in disparaging and vicious terms. Children being educated in the first decade of the 21st Century are subjected to a whole parade of half-truths and downright lies which are aimed at sullying any vestiges of glory that may be still attached to our imperial past.

There is something horribly Stalinist about all of this. It is a cynical rewrite of history, which, to my knowledge of least, is unparalleled in modern times - in Britain at least. Five or six months after I returned to the UK, I was talking to the teenage daughter of my then girlfriend. She was studying for a history GCSE, and I was doing my best to take an interest in what she did at school. She told me, indignantly, that the British had once engaged in the slave trade. Yes, I agreed. Sadly, they had. She asked me whether I knew that this was how black people had come to England? She went on to tell me that at school that day she had been told that Britain was a multicultural society because we had brought large numbers of black people to Britain as slaves.

Because she was only 15, and because I do my best to educate rather than rant and splutter, I politely - and gently, because this was her favourite teacher - told her that what she had been taught was complete nonsense. Yes, of course, we had been involved in the slave trade, but it had been illegal in Britain and her dominions overseas since the 1820s, and one of the only reasons for our imperial incursions into parts of West Africa later in the 19th Century, and in the early part of the 20th Century, had been in order to fight the slave trade.

A large proportion of the black people come in this country are descended from immigrants who came in on a boat called the *Empire Windrush* in the 1940s, only just about a decade before I was born, in order to fill the great gaps in the British labour market left after the carnage of the second world war, and the gaps in the social infrastructure left after the Irish Republic left the Commonwealth in 1948.

I don't think that she believed me. Her mother's hippy boyfriend did not carry anywhere near as much clout on an emotional level as her favourite teacher who coached the girls' hockey team.

I am not going to try to defend the British Empire entirely. We did some bloody appalling things - all you have to do is read the history of the Opium Wars to find that out. In more recent years, the way the we treated the Irish, and the inhabitants of the Chagos archipelago in what is now known as the British Indian Ocean territories, is indefensible, but on a whole - at least during the last hundred and 50 years - the British Empire was a pretty benevolent thing. However, I am one of the few lone voices outside the nutters of the extreme right-wing, who ever gets up and says so in public.

So, in my experience at least, in the 21st Century nobody likes to be reminded of their Imperial past.

Imagine my surprise, therefore, when I find out that the greatest cultural links that modern-day Puerto Ricans, and - if my new informants are to be believed - modern-day Cubans have, is with their one-time imperial masters in Spain.

Although Spain had not been the motherland for well over a century, somehow the Spanish colonial influence has continued. If they can afford it, Puerto Rican families will send their children back to Spain for at least part of their education. Many of the posher families on the island are as proud of their Spanish colonial ancestry as any member of the fast vanishing English gentry is proud of their genealogy. Indeed - I have found out later - being able to claim descendancy from one of Columbus's crew, is approximately as high a social cachet as being able to claim that your family came over with William the Conqueror, although in both cases, why anyone would be proud to claim that their ancestor was responsible - directly or indirectly - for centuries of genocide, is beyond me.

"Cheese" went on to tell me about the chequered history of the island up until the Spanish American War.

Spanish Colonial emphasis during the late 17th - 18th Centuries, focused on the more prosperous mainland territories, leaving the island impoverished of settlers. Because of concerns of threats from

European enemies, over the centuries various forts and walls, such as La Fortaleza, El Castillo San Felipe del Morro and El Castillo de San Cristóbal, were built to protect the port of San Juan. The French, Dutch, and English *all* made attempts to capture Puerto Rico, but failed to wrest long-term occupancy of the island.

In 1809, while Napoleon occupied the majority of the Iberian peninsula, a populist assembly based in Cádiz recognized Puerto Rico as an overseas province of Spain with the right to send representatives to the Spanish Court. The representative, Ramon Power y Giralt died soon after arriving in Spain. These constitutional reforms were reversed when autocratic monarchy was restored. Reforms in the 19th Century augmented the population and economy, and expanded the local character of the island. After the rapid gains of independence by the South and Central American states in the first part of the century, Puerto Rico and Cuba became the sole New World remnants of the large Spanish empire.

But as we have seen, even whilst the pan-European dominance of Napoleon Bonaparte would have given the inhabitants of Puerto Rico and Cuba the perfect chance to gain at least a measure of self-determination, they did not avail themselves of it. They were spiritually wedded to the Iberian motherland, and it appears that they still are.

However, together with this unlikely reverence towards their one-time colonial masters, the spectre of slavery casts a massive, and gloomy, shadow. This is very different to the modernist liberal bullshit which is spouted by the current British establishment and their lackeys - this is genuine, heartfelt, and with a very real basis for it. Many of the inhabitants of Puerto Rico have slaves well within their ancestry, and the barraccoons or slave pens can still be seen. Slavery was a disgusting trade, and one that the human race - even now - should exhibit prominently in our hall of shame. However, I think that - when you have seen, as I have seen - the very real historical scars of this shameful period of history, emblazoned across the landscape and infrastructure of an entire community, it shows the recent paltry displays of fake emotion, by our current government here in the UK, up as being about as genuine as a prostitute's smile.

Although the people of Puerto Rico are surprisingly proud of their heritage as part of the quondam Spanish Empire, they're also disgusted and ashamed both of the history of slavery, and of how the proud conquistadors from the motherland in Spain destroyed the indigenous Taino people.

In recent years the United States has started to exhibit a cultural conscience; and - being the United States - they have gone totally over the top about it! When I as a child, back in the halcyon and carefree days of the 1960s, cultural matters were much clearer. Red Indians were the bad guys, unless of course they had attached themselves to a white benefactor (usually a cowboy of some sort), and whilst - even on my favourite television programmes - the Red Indians had had a sort of noble savagery (OK, they got to tie girls to trees, and wear cool hats with feathers in), they were still somehow second best to the intrepid cowboy who got to wave guns around and drink two fingers of whisky in a dirty glass.

Then in the Seventies it all changed. The truth began to come out, and the television factories from Hollywood no longer treated "injuns" as convenient foils to their cowboy superstars. Some semblance of the truth - that the European immigrants to the New World had committed a shameful and revolting programme of genocide against the indigenous inhabitants of the Americas - began to appear on both the big and the small screens.

Something similar happened with cinematic portrayals of the Vietnam War, and - more recently - in portrayals of the Second World War. Things were no longer seen clearly in black and white, and the moral grey areas, which every adult recognises to be the truth about life in general, and war in particular, began to proliferate across the entertainment industry. However, it didn't end there.

Some time in the late 1980s the Red Indian was rebranded once more. This time, the Mighty Redskin became a new-age icon. Over the past 50 years, as adherence to conventional religions has waned, there have been several disturbing results. As I showed in my 1999 book the *Rising of the Moon*, one of the most significant growth areas in contemporary religion are the blinkered fundamentalists who rely on fomenting hysteria amongst their adherents. As I wrote nearly 10 years ago, there is - in my opinion at least - a burgeoning amount of evidence to suggest that some fortean phenomena can be linked with this unhealthy rise in the fundamentalist bullshit.

Another social movement that has grown apace over the past 20 years is the rise of the new-age spirituality. Basically, I believe that this is 'Religion-Lite'; a collection of esoteric nonsense from all over the world which is gathered together into a undeniably attractive, but ultimately vacuous whole, suitable only for the gullible, and for those who are not prepared to put in enough effort to follow an established religion. These people lapped up the new gospel of Native Americans spirituality as if it was cocaine at a 1980s dinner-party.

Now, don't get me wrong. I know a lot of people who have been beneficially influenced by an interest in Native American spirituality. But I also know a helluva lot more who have been suckered into buying vegan dreamcatchers, bison burgers made of tofu, and silly hats covered in feathers and sacred crystals. By writing this, I do not want to alienate the people who have made a serious study of such things, but I'm sure that I will not do so. The people who are serious about a belief system based on Native American spirituality are, I suspect, as irritated by the commercial bullshit merchants as most Christians that I know are irritated by the vacuous directions towards which the Church of England seems to be taking us.

But I digress. The important thing about this syndrome of cultural guilt is that whereas 40 years ago, the Native American was perceived as a brutal - if sometimes noble - savage, nowadays he is seen as an icon of new-age spirituality. To anybody with any degree of hindsight, it is obvious that neither view is more than slightly correct.

However, back in the bar of the Windchimes Hotel in Puerto Rico, the conversation was beginning to take a familiar term. Because, as well as their pride in their Spanish ancestry, and their overwhelming guilt for several hundred years of the slave trade, there is also the strangely edenic feeling of loss for the Taino people.

The Tanio, and in particular the tribes who lived on Puerto Rico, who called themselves the Boriquen, were a strange and spiritual people - at least according to current ways of thinking.

Whether they actually *were* any stranger than any other primitive peoples is open to debate, although the people I spoke to that night certainly believed that the Taino had achieved much higher levels of spiritual awareness than many other relatively unsophisticated peoples around the world.

On thing is certain, however: they were not savages. A typical Taino village was based around a central plaza, which was used for various social activities such as games, festivals, religious and public ceremonies. Unusually for a relatively primitive people, they seemed to have a degree of sexual equality, which is almost unprecedented. The chiefs could be either male or female, and although the practice of polygamy was widespread, both men and women could have multiple spouses. They practised a mainly agrarian lifestyle, but they also fished and hunted. They made jewellery out of gold and seashells, and their art was room surprisingly complex. Their society was divided into two classes: *naborias* and *nitainos* with the latter being their aristocracy. The third caste were the healers and priests known as *bohiques*, who are still famous for their healing powers, and their alleged inability to speak with the Gods.

Taino shamans used a number of instruments in their healing practice. These included stone pestles to **prepare** the cohoba, vomiting sticks used for ritual purification, vessels to store the psychedelic ground seed of the cohoba tree, and forked tubes used to inhale the cohoba. Many of these objects are decorated with bat and owl motifs, probable symbols of death for the Taino.

The theology of the Taino was complex, but had something notably in common with one of the best-known pantheons of European gods. [1]

The Creator God, analogous with the Alpha of Judaeo-Christian theology, and Zeus of the ancient Greeks was called Yaya. He killed his son Yayael for reasons that remain obscure, and it is said - in one of the many Creation myths - that Yayael's bones became the fish that swim in the sacred waters.

Another important figure is Itiby Cahubaba - the Earth Mother. It was her image which reminded me so strongly of the Celtic *sheela na gig* which I now realise that I saw so many years ago when Graham and I explored the cave systems above Aguas Buenas.

As Maria Poviones-Bishop [2] wrote in 2001:

"In the second or alternative form of the myth [of Yaya and Yayael], Yaya goes to the fields and leaves the gourd with Yayael's bones hanging from the roof. Shortly after, four brothers arrive at Yaya's house. These four brothers are considered quadruplets, "all from one womb and identical." The myth says they were taken out by a type of "caesarean section" from their dead mother, Itiba Cahubaba, who died in the process of childbirth. The brothers begin to eat the fish in the gourd until they hear Yaya approaching. In haste, they try to hang the gourd back but are not able to secure it. The gourd falls to the earth where it breaks and lets out enough water and fish to create the earth's oceans.

With the introduction of Itiba and her four sons, the myth now appears to utilize a combination of metaphors for expressing processes of creation: the birth of new life as childbirth (in the emptying of the gourd and in Itiba's labor yielding four sons) and new life arising from death (fish created from the bones of the dead Yayael and live children born from the dead Itiba Cahubaba). The complex inter-play of these two ideas supports the hypothesis that the Taino saw death as a pre-cursor to birth, fertility and the re-creation of life."

Atabey (or Attabeira) was considered the central goddess figure of Taino religion. She was described as the mother of Yocahu, the god that lives in heaven and has no male ancestor. At least two European chroniclers of the early 16th Century indicate that this "mother of God" deity was honoured in an important yearly festival.

The Gods of ancient Greece were said to live on Mount Olympus. The gods of the Taino lived on El Yunque. According to another Creation myth from Puerto Rico, in the beginning, all people in the world lived on a single mountain. A boy living on the mountain planted some seeds, and a beautiful forest grew up. One night two men were having a fight over ownership of a pumpkin which had grown in the forest.

---

1. Most of the information in this chapter came from "Cheese" but it was confirmed from many sources including *The Tainos: Rise and Decline of the People Who Greeted Columbus* by I Rouse (Yale University Press; New Ed edition (2 Aug 1993) **ISBN-13:** 978-0300056969
Cave of the Jagua: The Mythological World of the Tainos by Antonio M. Stevens-Arroyo (University of Scranton Press,U.S.; 2nd Ed edition (30 Jun 2006) **ISBN-13:** 978-1589661127

2. http://www.kislakfoundation.org/prize/200103.html *The Bat and the Guava: Life and Death in the Taino Worldview* by Maria Poviones-Bishop July 30th, 2001

It rolled down the mountain, split open, and the ocean with all its creatures spilled out. Thus the rest of the world was created.

Other stories tell how the streams which flowed from El Yunque down to the sea, were magickal, and how all the animals in the world were created from the sacred waters. These stories were handed down from father to son amongst the bohiques, and it was undoubtedly one of the last of these tribal shamen who told these stories to Ponce de León.

The Taino believed that the dead went to a place called Coaybay, which was located on one side of an island called Soraya, which can be transtlated as 'sunset'. They believed that both the living and the dead had a spirit. For the dead, this spirit was called op'a. The op'a were believed to hide away during the day and come out at night to eat guayaba (guava fruit). The op'a were known to attempt to seduce living people, and op'a women were said to vanish into thin air when living men tried to put their arms around them. In addition, the op'a - who could transform themselves into a multitude of shapes, including fruit, or living relatives of the person to whom they appeared - came out at night to "celebrate and accompany the living."

The Taino believed that there was one way to tell apart the op'a from the living and that was by looking for the person's navel. The Taino believed the dead have no navels - especially meaningful if one considers that the navel is the point at which newborns are attached to their mothers. In light of the matrilineal descent customs of the Taino, the navel or physical link to the mother also determined a person's place in the community or society.

In many examples of Taino art, skull-like images alluding to death dramatically fuse with images of bats and owls, animals of nocturnal habits and sinister appearance that are associated with the opias. It is important to consider that bats are flying mammals. If the Taino did associate the bat with death, they chose a curious animal that both mimics humans in early life (i.e. bats nurse one offspring at a time) and demonstrates the distinctly non-human ability to rise up in flight. The latter is consistent with Taino notions of the op'a as beings that are not bound or tied to this life in the same way as the living goeiza.

For the Taino, from what we have seen, death was not extinction, punishment or reward. It was an episode in the transition from one existence to the other, an event expected and anticipated in the natural cosmic order. The dead were not in heaven or hell, or with the creator. They were on one side of the island, waiting for the sun to go down to come out and eat *guayaba* (guava), have sex, celebrate and dance.

And then it was over. "Cheese" and the boys embraced us, said their farewells, and disappeared out into the night.

Next to go was Nick Redfern. He had an early flight, and so with a *"Ta Ta Johnny, bostin loike"* he was gone. I was too wired to sleep with any degree of ease, and my days of chemical mood enhancement were long gone. So, as David and Kevin were off in search of a nice restaurant, I decided to tag along, and a very convivial and pleasant evening was had by all.

I don't know what it was about us, but we must have been obviously media types, which is something I find mildly reprehensible, as I have done my best all my adult life to be nothing of the sort. However, the ridiculously sexy waitress at the ridiculously expensive restaurant at which we ended up obviously thought that we were her one-way ticket to Hollywood, and sashayed around the table showing off as much of her cleavage as she dared. She was a singer, and an actress, she told us while pouting wildly, and she would do just about anything to get into the movies. I don't think she believed me when I explained that I was merely an impoverished zoologist from England, and whereas my two companions were indeed film-makers with an impressive pedigree, they specialised in documentaries.

I am sure that if I had been a less moral person I would have got lucky that night, but every man has his limits, and she was not only far beyond all of mine, but was so lissom that she would probably have given me a heart attack. But my adventures in the seamier side of showbiz were not over. Just as we were leaving the restaurant, another ridiculously attractive young woman leapt out at me, bared her breasts, and asked me to autograph them! Over the years I have had a number of strange things happen to me, but I don't think any of them compare even slightly with this. It transpired that she had mistaken me for notorious documentary filmmaker Michael Moore, but quite why she supposed that the man behind *Farenheit 9-11* would want to scrawl upon her mammary glands with a magic marker remains somewhat obscure. With my honour just about intact, and I got back into a taxi with David and Kevin and went back to the hotel. We said our goodbyes, and they went to bed. But I still had one more duty to perform.

My father was very ill and in his late seventies. My mother had died two years before, and - stubborn to the last - he had refused to have anyone live with him to look after him. Eventually - a year later - I moved in to the old family home, together with most of the boys from the CFZ, to look after him. But that is another story.

My father and I had a particularly difficult relationship. Neither of us had liked the other very much, and for most of my adult life we had done our best to pretend that the other one didn't exist. However, following my mother's death, I had taken it upon myself to telephone him every day, and the fact that I was in Puerto Rico didn't make any difference. Negotiating the maze of international phone calls from my credit card was a laborious and expensive business, but I did it, and the poor old chap was pathetically grateful for five minutes of a transatlantic telephone call. Towards the end of his life his relationship with me thawed, and he took an interest in my various activities. I told him about the events of the day, and promised that I would visit him as soon as possible upon my return. Despite everything, I loved him very much, and the idea of going to bed without saying good night to him was unthinkable.

Before I went to sleep, I packed my bags, fed my precious snails, and gulped down my medication. Tomorrow was going to be a long day.

The next day I arose early, and took my bags in each hand as I walked down the narrow street that led from the annexe where I had lived for the last few days, to the main body of the hotel. I had my last sumptuous breakfast, drank my last cup of coffee, and paid my bar bill. I was just sitting in the open courtyard watching the tiny yellow birds going about their busy affairs when Carola arrived. She was as scatty, and as sexy as ever, and I felt a real pang of heartache at our imminent parting. She gave me a gift; carefully wrapped in brightly coloured paper festooned with ribbons. I unwrapped it - it was a *Vejigantes* mask. *"So, Jon"*, she said with a tearful smile. *"You have a chupacabras after all"*.

I kissed her, and thanked her, and there were tears in my eyes as well.

On Graham's and my first trip to the island in 1998, we had both discovered a latent taste for honey rum, and as there were several hours to go before I had to check in, I asked Carola whether she could take me to the supermarket, so I could buy some to take back with me. Unfortunately, she misunderstood. She did indeed take me to a supermarket, but it was a very upmarket affair; a Latino version of Sainsbury's, rather than the Puerto Rican version of a seedy corner shop that I had been hoping for. The only honey rum that I was able to buy was made by Bacardi, and although it was quite pleasant, it was no way near as nice as the rough stuff which the winos drink, and which Graham and I had enjoyed so much. I'm sure that she was doing her best to show me the civilised side of island life, but - sadly - when I set my mind to it, I have deliciously unrefined tastes. She dropped me at the airport, kissing me goodbye, and then she was gone.

I checked my baggage in, and - in those carefree days before the war on terror - thought nothing of carrying my precious snails in my hand luggage. I decided to ignore the signs that were everywhere, exhorting potential passengers to be careful not to carry biohazardous material on to the planes, and I was soon in the air, heading to New York.

The Class of 2004

L-R Kevin, David, Carola, "Cheese", the author

The Margarita Diary,
Chapter 7 Comments:

One of the things that stood out for me upon reading this particular chapter was how, on any given CFZ adventure, the initial quest inevitably becomes something very different – and particularly so when new, and unforeseen, factors come into play. I'm talking here about the increasing accounts of something crashing into the El Yunque rain-forest decades earlier; Cheese's strange story of isolated pockets of human beings with progeria-style conditions; and tales of mutilated peacocks.

We had flown to Puerto Rico with the intention of trying to determine, for the benefit of the Sci-Fi Channel, if we could find, examine and identify any evidence for the existence of the chupacabras – such as undeniable DNA. Yet we were now deeply immersed in stories of genetic mutation, bizarre changes in the island's ecology and much more.

I well recall saying as much to Jon as I gainfully manoeuvred our jeep up what seemed like a near-90-degree slope, and through that huge herd of cows – a process that David had us repeat for the camera seemingly endlessly.

But for me, all of that was – and indeed still is – a big part of why I do the things I do: namely, the sheer entertaining unpredictability of not knowing what might come next, of finding something different and unexpected around the corner, and of heading off in search of new clues, new sources, and new leads.

As Jon also notes: as this aspect of our quest drew to a close, so did my time on the island of paradise: I was due to fly back to the United States a day before Jon headed back to England. Thus, after completing what I like to call 'Project Peacock', we all had a final dinner and drinks in San Juan, toasted to friendships old and new, quaffed endlessly to a job well done, and said our farewells.

Jon aside, of course, I have never seen the others again: they are today merely the stuff of memories and faces on photographs. However, I will never forget that week in the summer of 2004 when Jon and I roamed Puerto Rico's rain-forest, its lowlands and its little villages in search of monsters. It was an experience that will stay with me for all my life, and one that was as much about friendship, adventures and good times as it was about hunting the chupacabras. And at the end of the day, that was good enough for me.

And with that all said, and with my part in the story now firmly at an end, I leave you solely in the more than capable hands of the good Mr. Downes. In the final two chapters of *Island of Paradise*, he reveals to you his intriguing, thought-provoking and unique theories on what may ultimately lie at the heart of the mystery of the Goat-Sucker – and while not forgetting his sterling, 007-like adventures as a transatlantic snail smuggler.

Adios, my friends! And enjoy the rest of the ride! **NR**

Chapter Eight
The boys of the old brigade

I have no idea why New York is referred to as the 'Big Apple', and be honest I don't really care. I have only ever been there twice - on my journey from London to Puerto Rico, and on the journey back. As I told a few chapters ago, my sojourn in New York on the outward journey was benign enough, although - as I was going to find out a few months later - it was eventful enough in its own way.

The journey back, however, was terrifying. It was only when I arrived on the tarmac at JFK that I realised quite what an appalling crime I had committed. For everywhere I looked there seemed to be signs warning of spectacularly nasty punishments to be meted out to those foolhardy enough to have imported a proscribed substance, or object, into the United States. Right at the top of that list of banned substances - alongside firearms, chemical explosives, and scag, were snails.

Potentially, at least, I was in big trouble.

It was election year in the land of the free, and the Great Satan himself was seeking re-election. The night before, in a ridiculously expensive Puerto Rican restaurant, a young nubile, free with her favours, had misidentified me as Michael Moore. God knows why. I am taller, fatter, madder, and with far more hair than Mr Moore, but for that year at least he was an American hero.

He had written a book called *Dude, where's my country?* which actually wasn't terribly good. However, it was written with the best possible intentions; namely to cause as much trouble for the incumbent President George W Bush as possible, and to limit his chances of re-election. Everywhere you looked that summer, there were piles of his book for sale. I found it quite surprising. Like every God-fearing anarchist ex-hippy I had been brought up with the cardinal dictat that America had an oppressive regime, and that although the British parliamentary democracy wasn't any great shakes, it was far better than the alternative from across the herring pond.

I still believe that this statement is vaguely true, but I cannot explain why this vehemently anti-Bush diatribe was so widely available. In fact - I have to admit - that I can't think of a single book ever published in Britain that was so stridently anti-establishment, and available on every station bookstall, airport news-stand, and in every shop window. The nearest one would have to a book like this in the UK, would be something like *Bash the Rich* by Ian Bone, or one of Penny Rimbaud's diatribes about Wally Hope and against Margaret Thatcher. Okay, both of these books, together with many other such libertarian, anarchic, and revolutionary texts can be found on my bookshelf, but I bought them in obscure backstreet hippy shops, or on the internet, and such things have certainly never graced the shelves of W.H Smith's.

Seeing piles of these books everywhere was quite a heartening experience for those of us whose view of the US government was basically a only a few opinion points lower than that of Stalin. However, it may just have been my paranoia, but there was a very nasty feeling around New York that summer's evening. There were armed police, and troops everywhere. It almost certainly *was* my paranoia, but most of them seemed to be looking straight at me.

I probably wasn't a very prepossessing sight. For some reason the water supply to my hotel room had packed up that morning, and so neither the snails nor I had enjoyed our customary morning shower. Puerto Rico is very hot, also as I have written elsewhere in this narrative, and I had been sweating like a proverbial pig. New York seemed even hotter, and I was forcibly reminded of those immortal lines from John Sebastian about summer in the city when the back of your neck "feels awfully gritty".

Although I didn't know it at the time, I had just become a diabetic, which made me perspire even more. I'm 6 ft 7 in height, weigh 24 stone, and have long hair over my shoulders. I was also wearing a t-shirt, which was emblazoned with the message: *"Anarchy, Peace and Freedom"*, which I suppose - with hindsight - was not a particularly good idea. Especially, it turned out, as New York was in the midst of a major terrorist alert.

Bloody hell! What was I going to do?

It was simple. I had three options. I could dump my snails somewhere, and leave them to die a slow and miserable death. I could surrender to the forces of law and order, and throw myself upon their mercy. If I did this, I would certainly be arrested, probably imprisoned, and my life would be particularly unpleasant for a few years - and the snails would be killed. Or I could do what I had already decided that I was going to do, and bluff my way through, hoping that I would get away with it.

This is what I did, and much to my amazement, I did it - indeed - get away with it. There was a particularly awkward moment in the departure lounge. I was sitting, doing my best to be as unobtrusive as possible, though as I looked like one of the less salubrious members of the Manson family, and stank to high heaven, the chances of my being unobtrusive were next to zero. About an hour before we were due to board our flight, a bevy of armed police stormed into the departure lounge and started to search people's bags - apparently at random. Here goes, I thought. I've had it now.

But they ignored me. An apparently innocent looking older lady with a blue rinse who'd been sitting three or four seats away from me, was led away in handcuffs, and in tears. But the grossly overweight anarchist hippy with a bagful of snails was ignored, and - an hour or so later - we boarded the plane and took-off.

I breathed an audible sigh of relief. I knew that I'd be safe landing in Britain. For one thing, although taking snails into mainland USA is tantamount to going up to an immigration official and asking for a one-way ticket to Guantamamo Bay, bringing them into Britain was not an offence (although having taken them without permission on to a transatlantic aircraft probably was). Once we had landed, I knew I was safe. I knew that the only people who get pulled over by customs are those who look like they are smuggling valuable goods, and I also knew that the British Customs and Excise officials are so apathetic that unless I been wearing a t-shirt sporting the motto "I've got two kilos of heroin in my bag" I would be unlikely to have been stopped.

Graham was there waiting for me, and after an uneventful journey my precious snails and I were back at my home in Exeter. For reasons that I will not bore you with in this narrative, I traversed the Atlantic Ocean more times between November 2003 and November 2004 than I have ever done at other times in my life. And I soon realised that trying to negotiate the rigours of British Rail after a swingeingly uncomfortable transatlantic flight, is just too much for body and soul to bear. Therefore - whenever possible - Graham is always there in the arrivals lounge at Gatwick to meet me.

I have to admit that I sometimes feel mildly guilty about this, because Graham really doesn't like early mornings, at the best of times, and having to get up at half-past four in the morning in order to meet my flight which arrives at nine, is a particularly irksome responsibility. However, he has family who live in one of the small suburb towns just north of London, and so he usually manages to combine a family visit with the responsibility of having to collect me from the airport.

The first thing we did as I arrived home was to unpack the snails, and to my great relief, 11 out of 13 that I had taken from San Juan the previous day had survived. I handed them over to Richard in his official capacity as zoological director of the CFZ and went upstairs to have a shower, and a brief lie-down before facing the rigours of what was left of the day.

As often turned out be the case, there's no such thing as a brief lie-down, and the next thing I knew it was about 8 o'clock at night; I had been asleep for about six hours. Marshalling resources of mental energy that I really didn't know that I had, I left that womb-like comfort of my bed, and went downstairs to see what was happening. Richard - God bless him - had set up two neat little 24 inch vivariums, which now each held half my precious snails.

They were basically two 2ft fish tanks, with glass lids, a heat source, and several inches of bark substrate, on which a selection of green vegetables were lying. Some of my darling little snails had already begun to tuck in, but others were being somewhat recalcitrant. In the end we found that the thing that they liked most were dandelions, with cucumber covered in grated cuttlefish bone for added calcium as a close second.

I suppose this is as good a time as any to have a few words about Richard - although he has hardly appeared at all in the book so far, he's an important figure in the last few chapters, and really needs an introduction.

I first heard from him in the mid-1990s when the letters that my ex-wife and I received from a Mr R Freeman of Nuneaton painted a portrait of an earnest young man who - like me - had been bitten by the cryptozoological bug at a relatively early age, and who was now preparing to start a course in Zoology at Leeds University. He seemed pleasant enough, although his handwriting was execrable, and after he sent us a cheque to pay for couple of year's subscription to our magazine, we exchanged letters occasionally.

A couple of years later, we arranged to meet him at the *Fortean Times* Unconvention in London in May 1996. I have to admit, that over 11 years later, as I sit here writing these words, I am finding me it incredibly difficult to summon up any memories of that weekend at all; my marriage was falling apart, my life was a complete mess, and I was probably drunk - I really can't remember.

I vaguely remember the stocky bloke dressed in black who came up to Alison and me, and talked enthusiastically about his projected expedition to Tasmania in search of the thylacine. We promised to keep in touch, and - I blush to admit - I had so many other things on my mind that I pretty well forgot about him almost immediately.

Three months later Alison left me, and the worst period of my life began. I had no idea that my marriage was in imminent danger of falling apart. We had been having a bad patch for some months, but there had been bad patches before, and I, at least, took the Till Death Us do Part bit seriously, and it never occurred to me that my wife did not. The events of 28th July, therefore, came as an appalling shock, but something for which I was equally unprepared, was the almost clinical carve-up of my friends, which seemed to mirror the almost cynical re-apportioning of our joint property. People that I had known for years suddenly stopped speaking to me, as the people who had been *our* friends suddenly became *her* friends or *my* friends.

Within weeks of my first wife demanding the divorce from me, my social life was as unrecognisable from before, as everything else. People whom I only had a passing acquaintanceship with, suddenly became friends, and by the end of the year the vast majority of the people in my life were folk who just simply hadn't been there a year before.

One of these was Richard Freeman. During my long, lonely evenings; where once I would have watched television, or chatted, with my wife and quasi-adopted daughter, I spent many hours on the telephone. I don't know whether it was because we discovered that I - like him - had an obsession with cryptozoology that went far deeper than most people's, whether we just found that we had quite a few similarities as far as a sense of humour, tastes in music, and fondness for wine, women and song, or whether - perhaps - Richard was just feeling sorry for me in the aftermath of my wife's decision to go off and shag my erstwhile keyboard player, but his telephone calls became more and more frequent, and before too many months had passed, we became pretty close friends.

It turned out that his luck with women was just about as good as mine, and as I staggered the pubs of Exeter, fairly - in hindsight at least - appallingly, lurching from one doomed relationship to another, in a way which - many years later - I realised had been portrayed perfectly in *The Simpsons*, in the episodes following Millhouse's parents divorce, Richard would phone me regularly, telling me enthusiastically of some young woman that he had met at university, who equally regularly would drop him like a hot potato a few days later.

As Christmas approached, I suggested that Richard might care to come down to Exeter to spend the second half of what is euphemistically described as the 'Festive Season' with Graham and me. He agreed enthusiastically; far *too* enthusiastically, as I commented to Graham at the time. It was only many years later that I realised that he had come down under a complete misconception. I lived in a house in a rather unprepossessing little red-brick housing estate called Holne Court. Richard had somehow managed to convince himself that I actually lived in the country, and that Holne Court was actually some sort of minor stately home, and that I probably had a butler - and certainly was the owner of acres of historic parkland.

It took only a few seconds after he arrived at my humble abode to realise the enormity of his mistaken preconceptions. Far from being the squire of my village, and living in the sort of gentrified luxury to which Richard had always had a penchant, the director of the Centre for Fortean Zoology was actually a middle-aged, mentally-ill fat bloke with no money, a burgeoning drink and drug habit, and a head full of impossible dreams. I also had no food in the house, because at that stage in my life, for some reason I was living almost entirely off cheese and onion quiches that had been deemed to be past their sell-by date. Richard hated quiche. I had a bottle or two of whisky and some diet Coke, but Richard hated whisky, and he couldn't abide Coke.

But despite all of that, we got on like a house on fire, and I was quite sad when - a few days after New Year - Richard returned to his grandparents' house in the West Midlands. At that time, despite the fact that my personal life was at its nadir, my professional life was actually the most successful that it has ever been. 1997 was the high point of British paranormal publishing. There were no less than eight different magazines on the subject published each month, and each month I would write at least two articles for each of them under various names. I was the most famous that I have ever been, but that was a summer that everyone was famous, and that all you had to do in order to achieve a level of public popularity which seems totally ludicrous with hindsight, was to come up with some half-baked theory or other, and persuade one of the increasingly harassed magazine editors to publish it. I managed to do this at least once a month, and - no longer encumbered by an unsuccessful marriage to a socially naive partner - I was beginning to make a mark for myself in Media Society.

By March 1997 I was beginning to be approached by television companies on a regular basis, and one-day, a nice little Jewish geezer from Channel 4 gave me a ring. His name was Jon Ronson, and he just been commissioned by Channel 4 to present a programme called *'For the Love Of...'* which was basically a forum for people whose interest in a particular subject bordered on obsessive compulsive disorder, to get together and get up in front of a television camera and hold forth for an hour. Did I, he asked diffidently, want to be involved? And could I suggest some other people who might be appropriate guests, on an episode dealing with cryptozoology?

Well, of course I could. The first person came to mind was Richard, but for some reason or other he was unable to make it when we filmed the pilot, and so I brought along two close friends, Chris Moiser, now co-owner of a small zoo in north Somerset in which my wife has shares, and Richard Muirhead, a bloke I have known since my childhood in Hong Kong, and another acquaintance of mine - also called Richard.

With hindsight (and I am only too aware that I'm using this phrase a lot in this memoir, but an awful lot of things about my life would have been very different if I *had* had the benefit of hindsight) this was not a good idea. This *particular* Richard was one of the proliferation of hobbyists who had entered the fortean fray in the wake of the enormously popular television programme the *X Files*. He was somewhere between a self-opinionated oaf, and a complete idiot. One of my favourite memories of him (if you can call something a favourite memory, when even now it makes me shudder with irritation) was when he telephoned me up in the middle of the night, and portentously told me *"We've got ourselves an EBE situation here, Jon"*.

I sat back in bed, stunned. What on earth was this appalling fellow talking about? What was 'An EBE Situation' and what on earth could it have to do with me? It turned out that 'EBE' was an unlovely acronym for extra-terrestrial biological entity, and that somehow this complete idiot had discovered that some self-styled researchers in Israel had got hold what they believed was an alien foetus. During 1997 there was an *idée fixe* amongst certain portions of the fortean community about alien foetuses. Why beings from a far distant galaxy would travel for aeons across uninhabited space, only to have a miscarriage in some remote village in Israel was beyond me. But this was only one of several reports of such things during that ludicrous year.

Anyhow, how did this socially deficient pork butcher from Finchley somehow come to the conclusion that it was his sole responsibility to co-ordinate international research into something that had happened in the Holy Land? God alone knows, but basically he was an idiot. I tried talking sense into him. How was he going to get hold of this aforementioned alien foetus, and what was he going to do with it once he got it? If, indeed some unfortunate female alien had decided upon carrying out this particularly unpleasant way of proving once and for all that aliens exist, surely the evidence should be in the hand of the world's scientists? Had he thought of contacting the British Museum, or the Smithsonian Institute? Of course not he said. They were all part of the International Military-Industrial complex who were bent on perpetuating a worldwide conspiracy aimed at making sure that Pork Butchers from Finchley were kept blissfully in the dark of the true facts; that aliens were - not-so - covertly manipulating world governments for their own ends. And the that the so-called foetus had landed in Israel, was ineffable proof - if any proof were needed - that the whole thing had been masterminded by the Jews. [1]

As an aside, I think I should mention here, that although - as you will find out in a few paragraphs - whatever quasi-professional relationship I had with him (in his own mind at any rate), was to end in tears before the verdant spring of 1997 blundered aimlessly into the long hot summer, I was to have one last contact from him, in mid-September 2001. After a self-imposed silence which had lasted four years, which had been pretty well enough for me to have forgotten all about him, he telephoned me late one night, to tell me that he had conclusive proof why the Jews had masterminded the horrific events 9-11. *"The proof is right there in front of us"*, he spluttered down the telephone. *"Why do you think the mayor of New York is called Jew-liani?"*

"Oh fuck off" I said, and put the telephone down.

But back in the early spring of 1997, the scales were yet to fall from my eyes, and I committed the potentially awful professional gaffe of inviting this paranoid idiot on to a national TV show. Luckily, my career survived!

Looking back over an interval of more than a decade, it is hard to try and recapture my mindset from those times. My darling wife often accuses me of being a narrow-minded old git who never makes more than the most cursory effort to understand the other person's point of view. She is known to quote the lines from her favourite novel - *To kill a Mockingbird* - about the desirability of being able to walk a while in somebody else's shoes. I freely admit that she's right, and that on the whole, when I find myself irritated at somebody, I have no interest in trying to examine the situation from *their* point of view. However, whilst writing this book, I have found myself in an even more peculiar situation; that of trying to get inside the emotions that I felt 10 years or more ago, with a view to trying to understand - and sometimes justify - my *own* actions.

The truth is, that the Jon Downes who staggered around London in the early months of 1997 has less in common with the Jon Downes who sits in his study, in the tumbledown 200-year-old cottage that he shares with his wife, and a motley collection of animals and men, than I feel comfortable having to admit. For the first time in my life I am getting embroiled in a generation gap....... with myself.

As I have written elsewhere, much to my surprise, in the spring of 1997 I suddenly became more famous than I have ever been since, but most importantly, far more famous than I had ever been before. It was well under a year since my wife left me, and my feelings of self-esteem which have never been particularly high, were probably at their lowest ever ebb. I was drinking too much, smoking too much dope, and I was an untreated manic-depressive. No wonder my critical faculties were not at their best.

Graham and I drove to London for what turned out to be somewhere between a screen test and an audition. With us in the car was Chris Moiser, a zoologist friend of mine who I had met for the first time a couple of years before during my ill-fated attempts to become an adult education tutor. The attempts weren't ill-fated, I was - and presumably still am - on the register of Devon County Council approved

---

1. For those of you interested in such things, I believe that the so-called `alien embryo` turned out to be a dead lizard that had somehow been dropped into a cow pat - presumably by Mossad, the Trilateral Commission and someone from the New World Order.

adult education tutors, but the sad truth was that hardly anybody within the group of people who sign up for evening classes within the City of Exeter have any interest whatsoever in cryptozoology. I taught for one term, to a class of four, and have never bothered to repeat the experiment. The jolly nice people at the Devon County Council adult education forums asked me to repeat the experiment for one day as part of their summer schools season during the school summer holidays of 1994. Reluctantly, I did so, and it was as I thought. Nobody was even slightly interested. Only one pupil turned up. His name was Chris Moiser.

Over the years we became close friends, and I decided to invite him on to this new TV show, because I knew that he was not only intelligent and articulate, but an interesting person. During the long, boring, and seemingly endless journey from Exeter to Battersea, we talked about mystery cats, sea monsters allegedly buried on a beach in Gambia, and various other cryptozoological cabbages and kings.

These days everybody has a satnav, and a mobile phone. It is practically impossible to get lost. Back in 1997 we had neither, and spent what seemed like hours travelling round and round the interminable backstreets of Battersea in search of the studio. When we found it, it turned out to be part of a semi-converted comprehensive school, which had none of the amenities that one would normally expect from a TV show allied to one of the four terrestrial channels.

Even though I had lived in Exeter for about 12 years at that time, I still consider myself to be a country boy at heart. So did Graham, and indeed - although it took best part of another decade to achieve it - we had always planned to move the CFZ into the country as soon as we could. The gloomy backstreets of Battersea were oddly menacing. They were redolent of inner-city decay. Although it was one of the first nice days of the year, and the sun was shining, the streets seemed dank and unwelcoming. All the buildings were built of red-brick, and there was a depressing post war homogenous feeling about them. Once upon a time the streets that we were driving through had been a hotchpotch of Victorian buildings, each with their own innate character, and each with a story to tell. The Luftwaffe had put paid to that and what we were driving through was a hastily constructed, and very cheaply built, domain for heroes which have been thrown together in the immediate post-war years, and which was now falling apart rapidly.

The streets were filthy, and mostly deserted. When we did see any other people, we seemed to be the only Caucasians around. I have often been accused of racism when I write such things, but I *do* feel uncomfortable when I am the only white face in a street full of people who - to my paranoid mind at least - appeared to be looking at me with jaundiced and suspicious eyes. However, the racist graffiti on the unforgiving brick walls, would tend to explain *why* the residents of the area were so suspicious of visiting white folks, and black friends of mine have told me how isolated and alone they feel in belligerently white areas, but that didn't make me feel any better.

Feeling ever more alienated, we were relieved when, at last, we found our destination.

Having parked the car in what had once been the playground, the three of us gingerly went inside the old school. This was probably not as peculiar an experience for Chris as it was for the rest of us, because he at the time was a teacher, albeit in a sixth-form college, but Graham and I hadn't been inside a school for 20 years, and although we knew perfectly well that we were not doing anything wrong, it still felt very much as if we were trespassing, and even more so that we had once again assumed the mantle of the naughty schoolboys that we had been two decades before.

I always remember from my childhood how strange it felt to go back into school either after hours, or in the holidays. Corridors which were perfectly familiar to me during the day time, suddenly seemed strange, and even forbidding when they weren't populated by hordes of screaming children. I had that same feeling again as the three of us tried to negotiate our way through the deserted corridors in search of a makeshift studio, which C4 had assembled in what had been once the assembly hall.

But eventually there we were, sitting around a long table drinking tea, and having make-up applied to our faces. It wasn't long before I realised quite what an error of professional judgement I had made.

Chris was doing just what I had always known that he would do. He was talking professionally, and in an animated fashion about the subjects that the producers of the programme wished to hear about. So was Graham, so were the other guests that I had provided from my voluminous address book. But Richard, whose surname shall ever remain un-revealed, was blurting out the most embarrassing tirade of pernicious nonsense that I have ever heard.

He was claiming that the Royal Family were reptilian aliens. He ranted about British big cats being the escapees from government genetic modification laboratories, and he claimed that the Loch Ness Monster was a living dinosaur that had been released into the lake by none other than Queen Victoria. Worst of all - especially considering that Jon Ronson, the host, is Jewish - was his tirade of vehemently anti-Semitic conspiracy theory nonsense. The Holocaust never happened. The Jews were responsible for the IRA bombing of Canary Wharf. Bigfoot was a Jewish hoax....... and so it went on.

I took the producer aside and apologised to him. To be quite honest, I thought that my TV career was over before it had begun, and was mightily surprised when I was told that as long as I got rid of the appalling Richard, I could have a second bite at the cherry. This time I telephoned Richard Freeman, and the rest is history.

The appalling Richard took the end of his TV career personally, and blamed it all on me. A few weeks later I received a bilious letter from a group calling themselves The League of Independent Researchers. I had never heard of any of them, but they included organisations such as 'The Anti Alien-Invasion Squad', ' The Paranormal Wolf pack' and the 'Little Poddington Ghost Hunters'.

They were appalled, they wrote, at my `unproffessional` (spelt wrong) attitude towards my research. I was obviously in league with the Government, and only in it for the money. None of the members of the league would ever have anything to do with me again. The appalling Richard was of the self-styled President and `Co-oordinater`. (Also spelt wrong)

I few weeks later I received another letter from him. This was a remarkable document, which was badly photocopied, and had been distributed to everybody that he had been in contact with. He had concluded his research, so he proudly crowed. Bigfoot, Jack the Ripper, UFOs, ghosts, alien abductions and crop circles were no longer a mystery. He had solved them all to his satisfaction, and was now leaving the world of paranormal research behind for good. Instead he was joining a gym, and intended to become a professional body builder.

And apart from the one occasion in the wake of September 11th, 2001, he has been as good as his word. I cannot say that I am heartbroken.

However my relationship with the other Richard - the Gothic student from Leeds University was becoming ever more solid every day. He was one of the undoubted stars of the version of *For the love of cryptozoology* which was eventually broadcast. I come over particularly badly in this show. Graham and I had got spectacularly drunk the night before, and I had a horrible hangover. My performance was less than inspiring.

Once again I find it particularly difficult to walk in my 1997 shoes. I don't know if I was ever an alcoholic, but I came pretty damn close. I drank far too much, and looking back from the perspective of the happily married man of 2007, who doesn't drink much at all now, I find myself embarrassed when I look at what I once was.

But there is no point in crying over spilt milk. What is done, is done, and I'm only revisiting those times for this book in order to explain the events which led me to make my two momentous excursions to Puerto Rico.

But Richard, Graham and I soon found ourselves to be firm friends, and in 2007 we still are. Over the spring and summer of 1997 we collaborated on half-a-dozen other projects, and when - for the second new year in a row - Richard came down to Exeter, we had a proposition for him. When he left university the following summer, would he care to join us as the third full-time member of the CFZ management team? We were overjoyed when he said yes, and in July, after a spectacularly surreal weekend at a pagan camp in Bridlington, which saw the head of Religious Education at a London comprehensive become doubly incontinent, Graham embark on a journey of alcoholic and sexual excess unparalleled even within the annals of the CFZ, me annoy a bevy of pagans by not taking their idiocies seriously (and singing words of Irish rebel songs to the pagan drum workshop), and Richard getting piles, we drove back to Exeter, and Richard became a full-time resident of Holne Court.

The CFZ would never have developed the way that it has if it hadn't been for the unique chemistry between Richard, Graham and myself. We all shared many common interests, and although - in many ways - we are very different people, from very different backgrounds, we are remarkably similar; especially in our senses of humour, and slightly surreal outlook on life.

Over the next few years we lurched from project to project, and worked on dozens of television programmes - mostly for foreign TV, and with very few exceptions eminently forgettable. The CFZ was so desperately under-funded that we did very little apart from publishing four issues of our journal each year, and working on West Country based cryptozoological investigations. Our work was mostly funded by writing articles for every magazine who would have us, and appearances on the aforementioned TV shows, and - although it I shudder to admit it now - when we got hold of a decent story, we tended to milk it for everything we could.

A prime example of this were the strange events that took place in the environs of an unofficial pet cemetery in the Haldon Hills just outside Exeter. This article by Graham and me appeared in a magazine called *Devon Life*:

Looming over the City of Exeter are the Haldon hills, a heavily wooded area with an unsavoury reputation. The forest was once home to Smokey Joe, perhaps Devon's most renowned tramp. Spurned in love he became a nature-loving recluse. Those of us who were children in the late 1970s well remember waving to him as we were driven past by our parents. He always waved back, and those who knew him remember him as a gentle loner who preferred the company of wild creatures to that of men. Not all of Haldon's inhabitants are quite so savoury, at least two unsolved murders have taken place here. Haldon could well be termed a window area - a place where many Fortean phenomena occur cheek-by-jowl.

Deep in the middle of the Haldon Hills is an area of woodland with an unofficial pet cemetery. It is a quiet and gentle place which has for many years been the last resting place of many beloved animal companions. Although it has actually been illegal to bury corpses on Forestry Commission land since the Environmental Protection Act of 1990, the powers that be turned a blind eye until, in the aftermath of the Foot and Mouth epidemic, it was closed for good by order of HM Forestry Commission. Many people are very upset by this decision and feelings are running high. Scott Grant, the Area Forester, told us that although it seems unlikely that the decision will be reversed, people can have access to the cemetery to tend existing graves, although it will not be possible to bury any more dead pets in the future.

For years we have been receiving strange stories from the cemetery and its environs. Not surprisingly animal ghosts abound. A witness who has asked not to be identified told us how, late one summer evening, he was driving with his family through the hills past the cemetery. Suddenly, a dog crossed the road in front of his car. He slowed to avoid hitting the hound (that appeared to be a golden Labrador) but the animal simply dematerialised in front of the shocked witnesses.

One night in April 1997 at about 11.00 p.m Yvonne Jackson was sitting in a car on the top of Haldon Hill with her boyfriend and another young man when they saw a semi circular object with 3 rows of lights; approximately 20 lights of various colours flickering and rotating. It was at an elevation of approximately 50 or 60 degrees and seemed to be hovering in the sky over the Teignmouth Estuary. A few months later she described it to us with the picturesque phrase that it was *"bobbling around"* and when we spoke to her months later she could only say that *"It was bloody enormous..."*

A friend of her boyfriend reported the sighting to the police, who said they'd had lots of reports of flickering lights. However, the police at Teignmouth were less helpful when we contacted them and could only say that reports of such incidents had to remain confidential. This is palpably untrue, because as we found out when contacting various other community policemen over the following months we discovered that in most cases they were quite happy to give us information, although in many cases they preferred the information to be 'off the record'.

The most notorious mystery surrounding the Haldon Pet Cemetery is that of the grotesque "Beast of Haldon". Over the years grave robbers have visited the cemetery to steal jewellery and other valuables buried with the animals. This is an unbelievably callous act even by human standards, but over the last decade or so this already eerie place has become even more unsettling. Graves have been disturbed by something other than sick people, something interested in flesh not trinkets. Some folk have caught glimpses of a strange animal stalking the cemetery at night. One such witness was Colin Yeo, a former police marksman. Whilst driving along one of the narrow back roads that transverse the Area one night he saw a huge dark-brown cat like animal bound out of the shadows and across the road in front of his car. It was over 6 feet long. On another occasion, whilst walking his dog, he disturbed the beast whilst it was in the act of digging up a body.

Another witness described something "like a small shaggy bear with a long tail" lurching through the woods behind the cemetery, and yet another man that we spoke to told us how he was walking his dog through the woods, a stones throw from the cemetery when he felt the eerie sensation of being watched. He looked around but there was no-one there. He carried on but the unsettling feeling wouldn't go away. Feeling decidedly unsettled he was making his way back towards his car when he turned around to call his dog. He saw a gaunt, dark, shape – apparently that of a huge animal – in the bushes a hundred yards or so behind him. To this day he is convinced that it was stalking him.

There is very little doubt that there are big cats of several exotic species living wild in the Devon countryside, but the creatures seen near the pet cemetery on Haldon seem to be something far more arcane. There is a possible zoological explanation. The "small shaggy bear with a long tail" sounds remarkably like a wolverine. This, the largest member of the weasel family, has been extinct in Britain since the Ice Age, but could a few isolated specimens have survived until the present day? It seems unlikely, but stranger things have happened. It is a very strange place anyway with ghost stories and UFO reports, and accounts of these lurching animal phantasms of the deep woods merely make them stranger. It will be interesting to see whether the closure of the pet cemetery has any affect on the stories which emerge from the Haldon Hills as the 21st Century progresses.

The pet cemetery was an undeniably spooky location, and for several years we exploited it for all it was worth. The scenario that we presented to the TV companies was always the same, although we refined it over time.

Basically we took the premise that the pet cemetery was an ideal place for a big cat, and so - on each occasion - we made a TV programme around our efforts to lure said big cat towards sand or camera traps. For those not in the know, a sand trap is, potentially at least, a very useful way of obtaining footprints from an unknown animal. Basically one get hold of a large bag of builder's sand. One spreads it across the desired area, several inches deep, and rakes it smooth. One then places efficacious bait into the middle of the sand, and hopes that - when the target creature ventures on to the sand in order to investigate the bait - it leaves footprints.

This was one of the two strategies, which we always pursued diligently for a procession of TV companies from across Europe, North America, and Japan. We got this down to a fine art: we found a builders' merchant that didn't mind being filmed on a regular basis, and we did a deal with a little private abattoir on Woodbury Common, who were happy - in return for a small consideration - to have us being filmed driving up to their gates, and collecting a noisome sample of animal entrails from out of the large odiferous metal bin which stood discreetly in the corner of their courtyard.

However, it was our other strategy that was most impressive. Acting on the theory that the animals that have been seen on the Haldon Hills were pumas, we devised another TV friendly scenario whereby we would visit the Dartmoor Wildlife Park about 40 miles down the road, and go into their puma cage to collect dung samples which - we hoped might attract other animals of the same species that might be wandering about mid-Devon.

Okay, it didn't actually work, in that it didn't actually attract any pumas. But it worked fine as a televisual artifice, and audiences around the world thrilled to the sight of the three intrepid boys from the CFZ apparently risking life and limb as they entered a cage with six wild pumas. What they were not to know was that the pumas were so tame, that on several occasions they would come up and lick our hands.

We owed most of this to our relationship with Ellis Dawe, founder - and then the owner - of Dartmoor Wildlife Park. He was a shocking old rogue, who was not averse to cutting as many corners as he could, but he was undeniably fond of the animals in his care, and we were undeniably fond of him. With a great shock of white hair and big white beard, I always thought that he only needed to lose a leg, to be the spitting image of how I had always imagined Long John Silver.

We got away with this for some years, but the more times we performed the charade, for more than we got heartily sick of it, and began to add our own little embellishments in order to amuse each other. One that sticks in my mind took place on one of the last occasions that we performed the saga of the pet cemetery - on this occasion for a Japanese TV crew from the Tokyo Broadcast System [TBS]. I was arranging the oval pellets of puma faeces in one of the clearings in the wood. For my own amusement I was fashioning them into a faecal analogue of Stonehenge. Before I knew what was happening, a microphone had been thrust under my nose, and the undeniably cute Japanese girl who was presenting the show, and who only looked about 12, asked me what I was doing.

I looked up at her beautiful face. She was dressed completely inappropriately for a winter's day in rural Devon. She was certainly the only Oriental chick in the county wearing a very low cut leather bodice, a white PVC mini-skirt, and a matching pair of thigh-length boots. She looked almost unbelievably earnest. *"What are you doing?"* she repeated.

"I'm making a crap circle", I announced proudly, and my two compadres collapsed into helpless laughter.

But despite years of soul-destroying nonsense, during which we prostituted ourselves to any and every magazine or crappy TV show that would have us, things did progress, and starting in 2002 we had a string of high profile expeditions and investigations around the world. Bizarrely, we were victims of our

Island of Paradise

own success. As we became more successful, and achieved more, we saw less of each other.

A case in point was July 2004 as I returned from Puerto Rico. It had been my third foreign trip in less than six months. During May I had been in Illinois for nearly a month, and almost as soon as I returned, Richard went to Sumatra for the second time. We had only spent a few days together, before I flew away on my Antillean adventure.

I was looking forward to spending some time with Richard for the first time in some months.

I always thought that Richard should have gone to art school. In between his cryptozoological activities, which do - after all - take up most of his life, he carries out a string of quasi-situationist practical jokes that uneasily straddle the border between art projects, vandalism, and sheer inanity.

One of his projects, which carried on for several years, involved picking up the wooden spatula shaped sticks from ice-lollies and taking them home. Several brands of ice-lolly have fairly inane jokes, of the calibre of those usually found in Christmas crackers, emblazoned upon them. It was this that Richard decided to use to his advantage. He started writing his own - excessively stupid, and generally completely unfunny to all but the cognoscenti - jokes upon them. The idea was to send bundles of them to various unwitting recipients. I am not quite sure why.

One of Richard's and my greatest comedy heroes at the time was a bloke called Paul Rose. Those sharp-eyed individuals will see that he has written the introduction to this very volume.

He was at the time (1993-2003) the editor of the Channel 4 Teletext video game magazine, *Digitiser*. On the whole, the content of this magazine had very little to do with video gaming, and was basically a forum for Paul's peculiar sense of humour. This included jokes like:

> *Q What do you call a giant killer bat?*
> *A Super beast 47*
>
> *Q What are the ingredients of a shepherd's pie?*
> *A Sherriff's hair and poo*
>
> *Q What do you call a scientist with a white face and dark brown fingers?*
> *A Doctor pudding*

These jokes are either the funniest thing you've ever heard, or complete nonsense, depending on your point of view. Also depending on your point of view, Paul Rose is either a genius, or a mildly annoying bloke who really should not be paid for writing such crap.

All I can say is that Paul Rose and his peculiar sense of humour got Richard and me through some very dark times. Many years later, Paul became a member of the CFZ, and a personal friend, but that is - like so much else that threatens to sneak in and take over this narrative - another story.

That Paul's jokes inspired Richard and me is undeniable. However - as in this case - the malign influence of Biffo was counter-productive, because his influence on Richard was so strong that he had spent much of the time when I was in Puerto Rico working on a magnum opus which he had dubbed *The bumper book of Joke Funnies*, which contained 40 or 50 inane jokes, illustrated with even more inane drawings.

These featured the exploits of characters such as 'The Boystings Children' (ten badly drawn youngsters who drank a type of milk called boystings, and talked nonsense).

SAMPLE JOKE:

Q How many Boystings children are there?
A 10

Another set of characters was the '10 Fathom Pirates' (two badly drawn monkeys in pirate costumes) and 'Yoc-yoc' (a vaguely sinister goblin, with a dandelion clock, and a hat that looked like a mushroom).

One might imagine that two of the world's leading cryptozoologists who have not actually seen each other for some time, and who had in the intervening weeks both carried out major cryptozoological investigations in foreign countries, would spend their first evening together talking shop, comparing each other's adventures, and formulating highbrow theories which they would then take pleasure in promulgating across the internet.

Not so. We went to a Chinese restaurant, ate too much, took a taxi home by way of an off licence where we purchased a large bottle of port, and spent the rest of the evening reading Richard's stupid book, and talking nonsense. It was all Paul Rose's fault.

There was plenty of time for work, we reasoned, and we would be able to have a good run of the serious work of comparing notes, writing up expedition reports, and promulgating our theories across the Internet on the following day.

We each retired to our separate bedrooms in a highly jolly state of mind, having consumed the best part of a litre of port between us. The future was looking very bright indeed.

We went to sleep happily, but not for long. Just after two in the morning we were woken by a phone call. Richard's father had died.

When my parents died - my mother in 2002, and my father in 2006 - it was the culmination, in each case, of a long, and debilitating illness. My mother was 79, and my father 81. Both had been in hospital, and neither death was unexpected. Richard's dad died quite suddenly in his mid-50s. Although he had a history of heart problems, the death was completely unexpected, and Richard was very shaken.

The very next morning he left Exeter for his home in the Midlands, and our in-depth examination of the results of our two expeditions never did take place. It is certainly unfair to blame poor old Biffo, but if he hadn't have inspired Richard to higher and more peculiar excesses of silliness, we might have had a proper debriefing session, and the events of the rest of the year might well have been different.

As it was, the first time that Richard actually heard about some of the events surrounding my sojourn in Puerto Rico, was over three years later, in the dying weeks of 2007, when he accidentally overheard me dictating these final chapters on to my laptop.

Richard was away for three weeks, organising his father's funeral and consoling his family. During those three weeks I was left to organise our forthcoming convention - the Weird Weekend - almost single-handed, and therefore was unable to carry out as many of the post expedition tasks as I would have wanted to, and therefore some of the information in these final chapters is not as complete as I would otherwise have wished.

Although I am a country boy at heart, I had - by that time - been living in a rather unprepossessing suburb of Exeter for nearly twenty years, and although I have never liked living there, Exwick and I had reached a sort of unwilling emotional stalemate with each other, and just about put up with each other's existence without too much complaining.. However, in the spring, and the early summer, even the ugly red-brick housing estate on which I lived, which even, when I first moved there in 1985 had seemed not only spectacularly ugly, but to have been designed by an architect with all the aesthetic sense of a retarded sewer rat, achieved a kind of surreal beauty. Every summer, for 12 years, a family of house martins had made their nest under the eaves of the house next door, and no matter how miserable or depressed I was, the sound of chirruping baby birds demanding their next meal with a vociferousness I was not to know for myself until I finally acquired a family of my own many years later, would always bring a smile to my eyes, and a touch of spring to my increasingly jaded heart.

But that summer, the house martins didn't arrive, and without Richard cheering me up with his good-hearted inanities, Exwick seemed lonelier and more forbidding than ever.

Lisa - the homeless girl, whom my first wife and I had 'adopted' back in 1990 - had always suffered from mental health problems. I had done my best to treat her as a daughter, and missed her terribly when she left with my wife. Five years later she had returned - emotionally, much the worse for wear - and from early 2001 I had once again tried to treat her as a daughter.

But she had a taste for drink and drugs, and - because I would not allow such things in the house - she started mixing with some very unsavoury company. What made it much worse, was that some of these people had once been so-called 'friends' of mine. One old hippy in particular used to encourage her to visit him, and - despite the fact he didn't use drugs himself - used to buy large quantities of various substances, and leave them around his house, so that she would get wasted and have sex with him.

Sadly, he was not unique.

I cut-off my relations with these people once I found out what they were really like, but the damage was done. Lisa would turn up at my house in the middle of the night paralytically drunk, screaming abuse, and accusing me of the most horrible things. I had no right to lay my bourgeois trip on her, she would scream (despite not knowing what the word bourgeois actually meant), and would continue to spout this hippy psychobabble parrot-fashion as she had been taught by the people who had once been my friends, and who were using it as a justification for their sexual abuse of her in return for drinks and drugs.

One night the police brought her back. Another, she attacked me, and after being restrained by Richard and Graham, was taken away by the police. I hoped they would section her; she was obviously developing schizophrenia, and needed to be under professional supervision. But we were living in the mid-years of Tony Blair's Britain, when the nightmare that had been promoted by Thatcher was approaching fruition. There was no care in the community. The Community just didn't care.

I called my doctor - my family GP for 20 years. He told me that was nothing he or I could do. She was not a danger to herself or to others, and as such would not be admitted into secure accommodation. The only way this could happen would be if she were homeless. But I wasn't prepared to throw my 'daughter' into the street. However, in February 2004, her behaviour had deteriorated to such an extent that I was given no other option. Graham and I engineered a situation whereby I would appear to make her homeless, and Exeter social services would be forced into finding her somewhere to live, and furthermore, somewhere where she could receive at least a modicum of care.

This was the hardest thing that I have ever done, but after several nights having found her wandering naked around the house, eating flaked fish food, and trying to start a fire on the kitchen floor, I had no other option.

I had no daughter, Richard was away for an indefinite period, due to his father's death. His grandparents were not getting any younger, and it seemed logical that at some point he would have to leave Exeter to go back and look after them. And the house martins had deserted me. I didn't want to live in Exeter any more.

Back in 2002 I had decided that it was time to clean up my act. I have been manic-depressive all my life, and for most of it I have self-medicated with alcohol and drugs. I began to wean myself off, and by the time that Lisa finally left me I was drug-free. I still drank, but when in August 2004 I was diagnosed as a diabetic, even that was taken from me, and although I still enjoy a drink or two, my days as a substance abuse and all the inherent lifestyle peculiarities which that entails were gone.

In the first few years after my wife had left me I lurched from doomed relationship to doomed relationship, but after the events of Christmas 2000 which are described in my book *Monster Hunter*, I'd given up on women, and from 2001 until the summer of 2004 I had been celibate. I decided that I had to deal with my other problems before even attempting to get into a relationship.

But by the summer of 2004 I was solvent, drug free, reasonably sober, and ridiculously horny, which explains why, after meeting a buxom single mother of four at the Weird Weekend, I felt deeply into bed with her. If she hadn't had four children things would have been much easier; I would probably have realised fairly quickly that we were ridiculously incompatible, we had different religions, politics, and in fact didn't agree on anything much that involved having clothes on, and that she was worst possible partner for me. But I found that I loved being an unofficial stepfather, and by the autumn I had left Exeter, and was living with her in a derelict farmhouse about 15 fifteen miles away.

But I am running ahead of myself here. My sex life was not the only thing to benefit from the events of the Weird Weekend; I also received a lot of fascinating information, which was to be of major help to me in my quest to solve the mysteries of the Island of Paradise.

The Weird Weekend is the largest regular cryptozoological conference in the English-speaking world. It is a fundraiser for the CFZ, and since 2000 has featured the cream of the world's cryptozoological researchers. Although the conference is predominantly cryptozoological in nature, it is not exclusively so. We try to make sure that the speakers include ones that are of interest to the people who attend, even if they are sometimes a little off topic.

The original mission statement for the event reads: *"For years we had - individually and collectively - been guests at various conventions and conferences around the country, and indeed around the world, and we decided that we wanted to do one of our own. However we wanted to do a conference with a difference.*

In the UK at least, the UFO Community, the Fortean Community, the Cryptozoological Community and the Pagan Community are four totally different things. We felt that this was a pity. There were and are people who regularly speak at conferences for each of these communities who have a lot of interesting things to impart to members of the other communities but until now never got the chance. We decided that as we had a foot in each of the camps we would do our best to ensure that the Weird Weekend would, each year, be a forum where people from all the different communities could meet, make friends and talk to each other. The first event was held in 2000, and annually since then, proving that our concept was a success beyond our wildest dreams...."

The event had very humble beginnings. The first event was held in a tumbledown Parish Hall in Exeter. It was attended by only about a dozen people, and made a loss of £75. The next two events were held at Exeter University in conjunction with the University Science Fiction Society, and although they were marginally more successful, they still only just about broke even, and the 2003 event - the one which, ironically, was our first success - very nearly didn't happen.

Interviewed by the *Exeter Express and Echo* in 2003, Richard Freeman and I admitted that we only went ahead with the plans for the 2003 event because friends talked us into it, and that we had already decided that the 2003 event would be the last.

However, a new venue (the 16th Century haunted pub *The Cowick Barton*), and a broader policy of guests was a success, and the fourth event made nearly £1,000 ensuring that the event would continue.

One of the 'friends' referred to above was Judith Jaa`far, who was then head of BUFORA – the British UFO Research Organisation, the longest standing and arguably the most respected UFO Research Organisation in the world. On the Saturday night of the Weird Weekend, the cognoscenti usually end up back at my place burglarising my liquor cabinet.

At Holne Court Graham and I have built a completely illegal, lean-to on the back of the house, which until we moved out in 2005 was one of our main reptile rooms. It was also where the hardcore party animals of Weird Weekend 2004 congregated to drink my whisky. Sat around the table were about half-a-dozen of us, including Judith and her fiancé, plus Dr Darren Naish - an old friend of mine, and one of Britain's leading young palaeontologists, and me. We will all admiring one of the tanks of Puerto Rican snails, which I had smuggled back to the UK just over a month before.

I told the story of the UFO crash to Judith, and suggested - semi tongue-in-cheek - that BUFORA might be interested in helping fund the return investigation. Not greatly to my surprise, they were not, though they were interested - and so was Darren.

I was talking generally about some of my other adventures on the Island of Paradise when he butted-in. Apparently, there are no such things as freshwater mantis shrimps. But there are, I insisted, because - as you already know - I saw some in a mountain stream high on El Yunque.

There are many different species. They range in size from only an inch or so - like the ones in this pool - to well over a foot, and it has been suggested that because this is such a successful hunting strategy that mantis shrimps are not necessarily closely related to other. They may be a group of disparate crustacea, which have merely *evolved* similar feeding strategies. They are found throughout the tropics, but no-one had realised you could get them in freshwater.

The idea that all the species of mantis shrimp are not closely related is not as strange as it might sound. The phenomenon of parallel evolution is a well-known one in the natural world. In Australia for example, unlike the rest of the world, placental mammals never really became established. Instead the more primitive marsupials held sway, and whereas non-Australia marsupials - the opossums of the New World - are relatively primitive, in Australia marsupials have evolved and filled every ecological niche that is filled by placental mammals in the rest of the world. There are marsupial moles, small rabbit-like potoroos and rock-wallabies, and a range of small carnivorous dasyurids even *look* like weasels, stoats and civet cats – their small predatory counterparts. The larger kangaroos take the place of grazing animals like deer and in recent historical times there were even large dog and cat-like carnivores, which some cryptozoologists still believe exist to the present day.

The idea of a feeding strategy evolving independently animals only distantly related to each other can be illustrated nicely in the case of sabre-toothed cats. Over the millennia at least two totally unrelated species of big cat, and one species of large carnivorous marsupial (which looked like a cat, although completely unrelated) had evolved with giant sabre teeth to attack their prey. Although there are no such creatures currently accepted by mainstream zoology, it would be a foolish man who would say that this highly successful feeding strategy would not lead to a new species of sabre-toothed carnivore evolving at some time during the future.

But, as far as mainstream zoology is concerned, there are no such things as freshwater mantis shrimps.

Darren also introduced me to the concept of microcontinents. [1] The huge tectonic plates, into which the crust of the planet is divided, are all based around a craton - a stable, Precambrian shield of rock that can be up to 3.8 billion years old. However, there are a number of microcontinents, which are built of continental crust, which do not contain a craton. It is believed that these microcontinental drift, following the same rules of continental drift which apply to the major tectonic plates. Several islands in the Caribbean have been proposed as microcontinents, and it has been suggested that Puerto Rico is one of these.

If this is true, and the Island of Paradise did not break off from mainland South America relatively recently, it means that it drifted across the Atlantic from Africa independently. This could be extremely important if, indeed, it turns out that certain aspects of its zoology have more in common with the old world, than with the new.

He also filled in the gaps in my knowledge of the historical mammals of Puerto Rico.

According to a list of mammal extinctions on the website of the University of Wisconsin, the following species have all become extinct on Puerto Rico since the arrival of the Spanish:

The greater Puerto Rican ground sloth *(Acratocnus major)*, and the lesser Puerto Rican ground sloth *(Acratocnus odontrigonus)* were both wiped out by the introduction of rats and pigs in about 1500. The former reached a weight of 150lb where the smaller of the two creatures was only a third of that. Megalonychid sloths were once widely distributed in the New World, including the West Indies; one still survives in the forest of South America (the "ai," or two-toed sloth, Choloepus). They varied greatly in size, from giants that weighed nearly a ton to species only slightly larger than the living one (about 10 lb). The lesser Puerto Rican sloth, at roughly 30 pounds, shows arboreal adaptations, suggesting it was partially adapted to tree-dwelling (even though it is often referred to as a "ground" sloth).

Fifty years after another mountain dweller, the Puerto Rican Agouti *(Hetropsmys insulins)* was wiped out, and in 1600, the Puerto Rican caviomorph *(Hextaxodon bidens)* - an animal similar to a degu - apparently left the face of the earth forever. Over the years, several other small mammals and at least three indigenous species of bat also became extinct. [2]

But for a time as not only the Island of Paradise; it was the island of the really strange - and often giant - rodents.

But could any of these animals, which after all became extinct very recently in geological terms at least, have survived to the present-day? And most importantly, could any of them be behind any of the stories

---

1. Darren told me this stuff, and a look at Wikipedia, and more importantly the references cited in their microcontinent artile confirmed it.
2. http://faculty.jsd.claremont.edu/dmcfarlane/extinctmammals/ Extinct mammals of the West Indies

of sightings of the Chupacabra.

Darren and I talked this over for some time. By the time we had finished our conversation everyone else had gone to bed. What Darren told me was enlightening.

Obviously the most exciting animals ever to have lived on Puerto Rico were the two species of ground sloth, and whilst it is certainly possible that one or other species has survived, neither is likely to be the true identity of the chupacabra, although certain aspects of the largest of the two species are somewhat reminiscent of the owl-faced creature that Norka reported seeing. However, the one thing that is usually forgotten by cryptozoologists who are keen to use a hypothetical surviving ground-sloth as a reasonable hypothesis for various South American cryptids, is that - by their very nature - they were as slow and lumbering as their living counterparts. The two Puerto Rican species were also, at least partly, arboreal, and have been dubbed as ground sloths more because of their size, than for any morphological reason.

However, Darren told me that he was vaguely aware of a long extinct Antillean porcupine, and he promised that he would find out more information for me over the next few months.

We said our good nights, and went to our respective beds.

I would like to say that I slept the sleep of the just, but Darren, Richard, and I were not the only people sleeping at Holne Court that night. Three or four other assorted luminaries of the fortean scene were also camped out at various - mostly inappropriate - locations around my house. One, a crop circle researcher who shall remain nameless, and who has - by the way - in the intervening years cleaned up his act, was at the time one of the biggest 'drug dustbins' that I have ever met; and that is saying something!

He had arrived at my abode halfway through Thursday, and - to the best of my knowledge - for the next four days didn't sleep a wink, mostly because he was refreshing himself at intervals from the contents of one of those little plastic cash bags that the high street banks give you to contain small change. However, this bag was not full of pennies and tuppences, but contained about half an ounce of white powder. Although by this time I was pretty well drug-free, I was still relatively tolerant of the peccadilloes of my peers, so - with mild amusement - I asked him what it was.

Despite my chequered past in this particular area, I was shocked by the answer. It was a mixture of LSD, Ecstasy, base MDMA, amphetamines, and ketamine! Basically, everything that he had in his house, ground-up in a pestle and mortar, and presented in an easy to ingest form.

Although I went to bed at about 4 o'clock, my crop circular friend decided to sit at the end of my bed talking what I believe in the trade is known as ketabollocks - the bizarre stream of consciousness free association which comes from the effect upon the cerebral cortex of unbridled ingestion of ketamine; a substance which is - after all - basically an animal tranquiliser.

Bloody hell, I thought to myself - not for the first time. I certainly have some weird friends.

The 2004 Weird Weekend was a great success, and made over a grand for CFZ funds.

But my life was just about to change completely. Within weeks I was living with my new girlfriend and her family in rural bliss, but as I have already alluded, the bliss did not last very long. By the time that I took her to America with me in November the cracks were already visible, and I was deeply unhappy. But the youngest boy called me 'Daddy', and I had a lovely relationship with the other three, and I couldn't bear to leave them. They had been let down by their father, and by my girlfriend's second husband, when she had kicked them out, and I didn't want it on my conscience to let them down for a third time. Christmas was coming, and although I couldn't see the relationship lasting much into the new year, I didn't want to screw-up her children's Christmas, so I sat tight.

However by mid-December I would have done anything to escape, and the universe - once again - gave me the perfect excuse to get away for a few days. I had an eMail from Manuel.

Manuel, you will remember, was the taciturn young environmentalist with an attitude problem whom I had met at JFK airport on my outward journey. I had sent him some details of CFZ projects by eMail soon after my return to the UK in July, and as I had heard nothing from him, I had pretty well forgotten all about his existence.

However, five months or so later he had contacted me again. What's more is that he was in London for a few weeks, and he wanted to meet up.

I loathe the metropolis.

Dr Johnson is supposed to have said that the man who's tired of London is tired of life. Well, I have enough problems keeping my suicidal tendencies under control as it is, and so I do my best to avoid London. But I was stuck in rural Mid Devon in a tumbledown cottage, which was freezing cold, which I was sharing with a young woman whom I was rapidly beginning to dislike. Manuel gave me the perfect excuse to get away for a few days, so - although I didn't think for a moment that the meeting would advance my researches - I jumped at the chance to meet him, and booked a return rail ticket for Paddington.

I dislike Christmas nearly as much as I dislike London. The birth of our Lord has been bastardised into the most disgustingly crass display of consumer capitalism that the world has ever seen. I know people who force themselves into near bankruptcy each year in order to celebrate a festival that they don't believe in, just because they have been socially conditioned to do so. I know people who spend literally thousands of pounds on presents, which are forgotten, and sometimes thrown away within days. How something so beautiful and wonderful, can be perverted into something so morally despicable is beyond me. Nearly everybody I know dislikes the season, and nearly everybody I know still spends money that they cannot afford to celebrate it.

Exeter station at Christmas time is particularly depressing. Christmas carols blare out of over the tinny speakers, the coffee shop is festooned with grubby tinsel, and most of the staff are drunk. The platforms were crammed with inebriated students on their way home for the holidays, and the words 'bah humbug' came to mind.

Al Stewart once wrote a song called *Trains* in which he told the events of twentieth-century history through the medium of ... you've guessed it: trains.

> *"In the sapling years of the post war world in an English market town*
> *I do believe we traveled in schoolboy blue the cap upon the crown*
> *And books on knee our faces pressed against the dusty railway carriage panes*
> *As all our lives went rolling on the clicking wheels of trains"*

A few verses later, the song continues:

> *"But oh what kind of trains are these that I never saw before*
> *Snatching up the refugees from the ghettoes of the war*

> *To stand confused with all their worldly goods*
> *beneath the watching guard's disdain*
> *As young and old go rolling on the clicking wheels of trains"*

And as I watched the heaving mass of humanity on platform five of Exeter St. David's Station, I hummed the song to myself, and told myself with a grim smile that the journey before me was going to be horrible!

When the train finally arrived my worst fears were realised.

There was only one spare seat on the train, and so I ended up sitting next to a buxom, stony faced, and slightly retarded looking wench - who, according to the name badge pinned precariously to her enormous, and somewhat shapeless bosom, was called 'Melena' - who not only smelt of an unwholesome mixture of gin and cheap deodorant, but was wearing an enormous pair of Angel's wings, fashioned artlessly from plastic and the sort of material that a third rate stripper would wrap around her breasts, in order to rip it away for her final dénouement. Above her head was an artifact of golden tinsel, which had obviously - at one time, at least - been intended to represent the halo of Angelic sanctity. However, because this young lady, who - I am sure - was completely unaware that her name was a medical term meaning 'blood in the faeces' was obviously having some difficulty in perambulation, her halo had been battered into a shapeless mess.

I hoped that my unwelcome companion would have the common decency to pass out before the journey was too much older, but I was to have no such luck. She spent the first 10 minutes of the journey chattering incessantly into a mobile phone to a friend of hers who was apparently called Chanterelle. It is not the act of a gentleman to listen in on a lady's telephone calls, but I just couldn't help myself. Anyway, she was conducting her conversation at such a loud volume, that to ignore her would have been nigh on impossible. Now, I am mildly embarrassed to admit that on occasions I swear like a trooper, and that I find nothing particularly embarrassing, or indeed reprehensible, about the use of the word 'fuck'. But I have never heard anything like the language used, or indeed the subject matters covered, in this excoriating conversation between a woman named after a blood soaked bowel-movement, and another one named after a mildly unusual wild fungus.

Apparently, Melena was a professional hostess (whatever that meant), and had been working in her professional capacity somewhere in Exeter the previous evening. Sadly for her, however, she had not been able to consummate her deepest feelings for a young man with the unlovely soubriquet of 'Big Bokka' (and yes, later in the conversation, she revealed - in excruciating detail - exactly why Bokka's name had been qualified by a size reference) because she was suffering from an unfortunate complaint in her nether regions. As the journey progressed, I - and, indeed, the whole carriage - heard about the aetiology of her condition, and by the time she staggered off the train at Reading, we would all have been able to have passed a *viva voce* examination upon the precise details of the colour, texture, and smell of her discharges.

A few seats down, a very drunk young man was pawing his unfortunate girlfriend, whilst swilling down Special Brew at a rate of knots. What made this public display of affection even more off-putting - and, indeed, bizarre - was that in front of them was a portable DVD player, which was playing *Gladiator* at full volume!

Behind me was a pair of tired-looking housewives, who were discussing their money worries. How, the most tired-looking one asked her friend, was she ever going to pay back the £3,000 she had borrowed in order to pay for her children's Christmas presents?

Island of Paradise

I did my best to fall asleep, but did not succeed in doing much more than drifting into a semi-lucid daydream. In my mind's eye I was back in Canóvanas, sharing the hospitality of my Puerto Rican friends. These were people who were happy with a plate of mofongo and red beans, and perhaps a glass of Medella. Although I had never been in Puerto Rico during what is euphemistically described as the festive season, I had arrived there are few days after Epiphany, and I had seen - though, sadly, never been able to share - the simple pleasures that these devout, gentle, and kind people seemed to take in celebrating that special time of year when the son of God was made flesh.

Eventually all good things come to an end, and - thankfully - so do bad ones, and after one of the most diabolical train journeys of my life, the rickety diesel engine, which has probably only a few years old, but like everything else in contemporary Britain was neglected, underfunded, and falling apart, pulled into Paddington Station.

I have to admit that I am quite fond of Paddington Station. It was built in the mid-19th Century by Isambard Kingdom Brunel, one of the greatest British architects. According to my family mythos, one of my male ancestors on my father's side, worked alongside Brunel in some capacity or other. My grandmother - who, admittedly, was prone to concocting stories of familial self-aggrandisement - claimed that her ancestor was actually the brains behind the operation, whereas my father - always one to place a wet blanket upon her more extravagant flights of fancy - claimed that he had been a mere navvy.

It has an undeniable air of imperial grandeur about it, and harking back once again to Al Stewart, at his song *Trains*, I have often wondered whether he was thinking about Paddington Station when he wrote:

> *"The silver rails spread far and wide through the nineteenth century*
> *Some straight and true, some serpentine from the cities to the sea*
> *And out of sight of those who rode in style there worked the military mind*
> *On through the night to plot and chart the twisting paths of trains"*

Paddington Station appeared in George Orwell's *1984*, a novel by Agatha Christie, and his reference in the movie *28 days Later*, but perhaps its most important contribution to the canons of modern culture came the year before I was born, when an undistinguished suburban couple called Mr and Mrs Brown were on the station waiting for their two children to come back from boarding school when they came across a specimen of *Tremarctos ornatus* (the only known South American ursid) carrying a carefully lettered cardboard sign reading: "Please look after this Bear!"

Paddington Station was even where - in March 2005 - I was to meet my lovely wife Corinna; but that is another story. The events of which I am writing, a few days before Christmas 2004, I was on my way to meet a terrorist.

I think that many people would have been terrified at the prospect. After all, terrorists are the social Antichrist for the modern world. As citizens of a so-called free Western democracy, I, together with all my readers in the UK, the United States, and Europe, are allegedly engaged in a titanic war on terror, which is - in fact - nothing of the sort. We are all ingrained to loathe and despise terrorists, and their sympathizers, but over the years - more by chance, than by intent - I have met quite a few, and several have even become personal friends.

I am sure that some members of the strange faction of folk who hide away on the internet, and seem to take great delight in taking pot shots at the CFZ in general, and me in particular, whenever we dare to put our heads above the parapet will be overjoyed by the previous paragraph, and flood the groves of cyberspace with allegations that me and the boys are hand-in-glove with Osama bin Laden, but I will have to disappointment them. But I know several animal-rights activists, including those who have taken the law into their own hands on a number of occasions. Whilst I was in Mexico, I met a number of members of the FZLN Zapatista separatist movement, and they seemed thoroughly nice chaps. And over the years I have met several members - and ex-members - of the various Irish republican paramilitary organisations, and at least two have become close personal friends. They have - with very few exceptions - turned out to be kind, gentle, retiring, and devout men who show no outward signs of the streak of psychopathy which one would imagine that someone capable of acts of violence on the scale propagated by organisations such as the Provisional IRA would have to possess.

I walked across the main concourse of Paddington Station, paused - as I always do - to say hello to the bronze statue of the nation's favourite bear, and then took the escalator to the first floor bar where I had arranged to meet Manuel.

There are two bars at Paddington Station; one – where, incidentally, my future wife and I had our first date - which is modern, bright, and a thing of polished chrome and post-modern functionality, and the other - *The Mad Bishop and Bear* - which is more traditional in nature, and - if you ignore the fact that for some reason the folk in charge insist on playing Bruce Springsteen records - could (with a little imagination) be the same pub in which Holmes and Watson must have sat, drinking porter, and smoking noxious cigars, before leaving London on one of their furtive expeditions.

It was probably because I knew that Manuel had never been to London before, and that I felt honour-bound as a red-blooded Englishman to share something of my country's cultural heritage with him, that I had arranged to meet him there.

He was sitting there waiting for me, and I have to admit that I hardly recognised him. Whereas when I had first met him, he had been dressed in a fairly nondescript manner, and looked like an earnest young librarian, albeit an earnest young librarian with wild looking eyes that oozed a suppressed anger, now he looked for all the world like a saint from an El Greco painting. The anger was gone, and appeared to have been replaced by a gentleness and wisdom beyond his years. His hair was longer, and he seemed to have found a new sense of purpose. He no longer looked like a revolutionary, and it seemed very difficult to reconcile this calm and spiritual young man with the knowledge that he had taken an active part in the Puerto Rican general strike of 1998, during which he had planted bombs, and helped to orchestrate widespread destruction.

We talked for hours about the history of the *Macheteros*.

The name *Machetero* is in memory of an impromptu band of Puerto Ricans who assembled to defend the town of Aibonito from the invading forces of the United States army during the Spanish American War, between August 10 and August 12, 1898. These *macheteros* did not formally side with the Spanish garrison that defended Aibonito (since, by then, most of its troops had either been wounded or had defected from the ranks), and assumed the town's defense on its own when the Spanish suffered heavy casualties. The Aibonito Macheteros successfully made the Americans retreat during a standoff sometimes referred to as the *Asomante Battle*, using a battle tactic copied from the Cuban *mambises* who fought the Cuban War for Independence in 1895 (charge the standing enemy with a rifle first, then slash him with a machete while he recharges).

He talked calmly about his hatred for the country that he perceived as a brutal and oppressive invader of his own land, and I was reminded of a similar conversation which I had almost five years earlier to the day with one of the perpetrators of the 1972 Aldershot bombing which killed seven people at the British army base, and was one of the last military actions by the official IRA before they disbanded.

This man too was an apparently gentle and spiritual soul, and talking to Manuel brought it all back. How could such a sweet, and apparently positive person be capable of such horrific acts?

I was reminded of the words of Derek Warfield, a much under-rated poet and lyricist, writing about the execution of James Connolly, one of the leaders of the 1916 Easter Uprising:

> *"They brought him from the prison hospital and to see him in that chair*
> *I swear his smile would, would far more quickly call a man to prayer*
> *Maybe, maybe I don't understand this thing that makes these rebels die*
> *Yet all men love freedom and the spring clear in the sky*
> *I wouldn't do this deed again for all that I hold by*
> *As I gazed down my rifle at his breast but then, then a soldier I.*
> *They say he was different, kindly too apart from all the rest.*
> *A lover of the poor-his wounds ill dressed.*
> *He faced us like a man who knew a greater pain*
> *Than blows or bullets ere the world began: died he in vain"*

Manuel, too had a smile which would have far more quickly called a man to prayer, but he too was prepared not only to die for his country, but to kill for it. He told me that he could never forgive the Americans for what they had done to his beautiful land, but that their two biggest sins were against the environment, and against - as he put it - the very hearts and souls of the people of the Island of Paradise.

He told me that what I had found out about the Monsanto testing of the family of herbicides of which Agent Orange was only a part, had only scratched the surface. The truth, he believed, was far more disturbing. Now, here I would like to stress that I have no evidence to support his allegations. On one level, and I suppose from a legal point of view, this can only be heresay from someone who - by his own admission - was a renegade from society, and who had committed terrorist acts against his lawful government. However, if you are to look at it like that, a large proportion of this book is heresay. As a journalist, and furthermore a journalist without large resources, or intensive financial backing, I have to take my sources where I find them. All the way through this book I have been entirely honest when I report what people have said to me. Some of them may well have been lying; indeed, on several occasions I have told you when I think that that was the case. But in this case, I would stake my reputation - such as it is that - Manuel at the very *least* believed in the complete truth of what he was telling me.

Wherever possible I have double-checked Manuel's assertions, and everything that I have been able to check, has proved to be true. Agent Orange had not only been tested in Puerto Rico, it had been developed there, and furthermore the history of the defoliant stretched back far longer than I had realised. According to Wikipedia (the free encyclopaedia):

"Agent Orange, given its name from the colour of the 55 U.S. gallon orange-striped barrels it was shipped in, is a (roughly) 1:1 mixture of two phenoxy herbicides in ester form, 2,4-dichlorophenoxyacetic acid (2,4-D) and 2,4,5-trichlorophenoxyacetic acid (2,4,5-T). These herbicides were developed during the 1940s by independent teams in England and the United States for use in controlling broad-leaf plants. Phenoxyl agents work by mimicking a plant growth hormone, indoleacetic acid (IAA). When sprayed on broad-leaf plants they induce rapid, uncontrolled growth, eventually defoliating them. When sprayed on crops such as wheat or corn, it selectively kills only the broad-leaf plants in the field, namely weeds, leaving the crop relatively unaffected. First introduced in 1946, these herbicides were in widespread use in agriculture by the middle of the 1950s and were first introduced in the agricultural farms of Aguadilla, Puerto Rico."

Manuel told me that the defoliant testing had been far more widespread than anyone had realised. Although Norka had told us how the American Government had leased a portion of the rainforest mountainside from her family for a period of 15 years starting in about 1960, Manuel claimed that this was a panic move on the part of the authorities, after a series of high-profile incidents during which local people had been injured and killed during effectively illegal experimental testing of both Rainbow Herbicides and biological weapons across the forests of the island. I can find no official verification that such tests indeed took place, and so if they did, they would have been effectively illegal, and done without official permission at governmental level.

According to Manuel, such official permission would never have been given, because - even in the late 1940s - the toxic effects of the chemicals within Agent Orange and its family of death dealing defoliants was well known. I have not been able to confirm this, but this extract from Wikipedia does suggest that by the time of the Vietnam War, these facts were known, even if they had not been made public:

"At the time Agent Orange was sold to the U.S. government for use in Vietnam, internal memos of its manufacturers reveal it was known that a dioxin, 2,3,7,8-tetrachlorodibenzo-para-dioxin (TCDD), is produced as a byproduct of the manufacture of 2,4,5-T, and was thus present in any of the herbicides that contained it. The National Toxicology Program has classified TCDD to be a human carcinogen, frequently associated with soft-tissue sarcoma, Non-Hodgkin's lymphoma, Hodgkin's disease and chronic lymphocytic leukemia (CLL). In a study by the Institute of Medicine, a link has been found between dioxin exposure and diabetes. Diseases with limited evidence of an association with Agent Orange are respiratory cancers, prostate cancer, multiple myeloma, Porphyria cutanea tarda (a type of skin disease), acute and subacute transient peripheral neuropathy, spina bifida, Type 2 diabetes, and acute myelogenous leukemia found only in the second or third generation. 2,4,5-T has since been banned for use in the U.S. and many other countries."

This article cites a 2000 paper published by the Institute of Medicine entitled *Veterans and Agent Orange: Herbicide/Dioxin Exposure and Type 2 Diabetes*. I have also discovered a paper published by the same organisation in 2004 entitled *Veterans and Agent Orange: Length of Presumptive Period for Association Between Exposure and Respiratory Cancer* which goes a long way towards substantiating both Wikipedia's and Manuel's claims.

Manuel told me that he had originally been just what he claimed - a science graduate with a deep and enduring love for the natural world. He had come from a military family - I knew this already, having met his father - and from what I could gather, his family had been quite an affluent one. As a child he had worked as a volunteer at the Luquillo Aviary, where - since 1968 - a recovery programme for the species had been organised by the Puerto Rican Department of Natural Resources. There he had fallen in love with the species, and decided to dedicate his life to trying to preserve it.

During high school, he had followed in his father's footsteps, and been a member of the Junior Reserve Officers' Training Corps (JROTC), a federal initiative founded in 1916, and - upon returning to Puerto Rico after going to University on the mainland, he joined the 92nd Infantry Brigade Combat Team of the Puerto Rican National Guard. As a part-time soldier, and full-time conservationist he spent much of his time in the forest, and slowly he began to become aware of the horrible truth that he believed was inherent in the United States occupation of the island.

One day when he was about 25, he was hiking alone across the foothills of El Yunque, when he came across an isolated campsite. That there he met a group of - what seemed on the face of it - to be dedicated environmentalists, and over a period of months he became friends with them. It was they who corrupted him (according to his father), or - in his words - taught him to see the world with his eyes open.

According to accepted wisdom, the Puerto Rican parrot became critically endangered after its habitats were substantially destroyed during the late 19th century. However, Manuel soon discovered that this wasn't quite true, or at least it wasn't the whole truth. During my first trip to the island in 1998 I had noticed that the ecology of the rainforest, at least in the areas where I had visited, seemed to be seriously

WANTED BY THE FBI

BANK ROBBERY, INTERSTATE FLIGHT—ARMED ROBBERY, THEFT FROM INTERSTATE SHIPMENT

VICTOR MANUEL GERENA

FBI NO. 134 852 CA2

Photograph taken 1982

Photograph taken 1983

Aliases: Victor Ortiz, Victor M. Gerena Ortiz

DESCRIPTION

Age: 25, born June 24, 1958, New York, New York
Height: 5'6" to 5'7"
Weight: 160 to 169
Build: Medium
Hair: Brown
Occupation: Security Guard
Remarks: Customarily wears light mustache
Social Security Number Used: 046-54-2581
NCIC: P0TTTT1016DIAA032212
Fingerprint Classification: 10 0 5 Tt 16 Ref: 13
                                I 17 A   17

Eyes: Green
Complexion: Dark
Race: White
Nationality: American (Puerto Rican descent)

CAUTION

GERENA IS BEING SOUGHT IN CONNECTION WITH THE ARMED ROBBERY OF APPROXIMATELY $7 MILLION FROM A SECURITY COMPANY. HE TOOK TWO SECURITY EMPLOYEES HOSTAGE AT GUNPOINT, HANDCUFFED, BOUND AND INJECTED THEM WITH AN UNKNOWN SUBSTANCE IN ORDER TO FURTHER DISABLE THEM. GERENA IS BELIEVED TO BE IN POSSESSION OF A .38 CALIBER SMITH AND WESSON REVOLVER AND SHOULD BE CONSIDERED ARMED AND DANGEROUS.

A Federal warrant was issued on September 13, 1983, at Hartford, Connecticut, charging Gerena with the crimes of bank robbery, unlawful interstate flight to avoid prosecution for armed robbery, and theft from interstate shipment (Title 18, U.S. Code, Sections 2113 (a) (d), 1073 and 659).

REWARD: A PRIVATE COMPANY HAS OFFERED UP TO A $250,000 REWARD FOR RECOVERY OF THE MONEY AND $100,000 REWARD FOR INFORMATION LEADING TO THE ARREST AND CONVICTION OF VICTOR MANUEL GERENA.

IF YOU HAVE ANY INFORMATION CONCERNING THIS PERSON, PLEASE CONTACT YOUR LOCAL FBI OFFICE. TELEPHONE NUMBERS AND ADDRESSES OF ALL FBI OFFICES LISTED ON BACK.

William H. Webster
DIRECTOR
FEDERAL BUREAU OF INVESTIGATION
UNITED STATES DEPARTMENT OF JUSTICE
WASHINGTON, D.C. 20535
TELEPHONE: 202 324-3000

Entered NCIC
Wanted Flyer 514
September 21, 1983

Ojeda released on $1 million bail.

depleted. Manuel and his new friends had noticed the same thing, but it was his new friends who informed him that - in their opinion at least - the main reason for the environmental degradation in so many areas of the island, had been almost entirely the result of repeated chemical and biological weapons tests, and the research and development behind the Rainbow Herbicides.

Manuel was a passionate young man, and in the same way that he had pledged his life to the protection of the Puerto Rican parrot when he was just a child, as a young adult he was finding a second crusade to join. However, it wasn't until he had known his new friends for nearly a year, that he discovered that they were all members of the Boricua Popular Army. It wasn't long before he joined them.

He travelled to Cuba - officially on vacation - but actually to meet Victor Manuel Gerena. Victor Manuel Gerena (born June 24, 1958) is an American linked by the Federal Bureau of Investigation with the armed robbery of a Wells Fargo armored car facility, in connection with *Los Macheteros* group. On May 14, 1984, he became the 386th fugitive to be placed on the FBI's Top Ten Most Wanted Fugitives list. He is still at large, and has spent the second longest time on the list since its inception in 1950. According to FBI investigations, Gerena was transported to Mexico, where he boarded a Cubana de Aviación jet at Mexico City International Airport in Mexico City, arriving at José Martí International Airport in Havana. [1]

If it hadn't been for "Cheese", on our last night in the bar of the Windchimes Hotel, telling me of the links between Puerto Rico and Cuba, I would have been surprise at the idea of somebody from an American-controlled island visiting the land of Fidel Castro so easily, but - in the light of what he told me, and what I have discovered since - Manuel's claims seemed perfectly reasonable.

Manuel was a veritable gift to the freedom fighters. He was a trusted, and well liked, member of the National Guard, and furthermore, because of his college education, he appeared to have a bright future within the organisation. Thus making perfect material for becoming a double agent.

However, Manuel was no fool, and did not take anything that he had been told at face value. He began to research the history of the relationship between the American armed forces and the island, and he was appalled at what he found.

Isla de Vieques, for example, is an island-municipality of Puerto Rico in the northeastern Caribbean. Although Puerto Rico is a U.S. Commonwealth, Vieques, like the rest of Puerto Rico, retains strong Spanish influences from 400 years of Spanish ownership. [2]

Vieques lies about 8 miles (13 km) to the east of the Puerto Rican mainland, and measures approximately 21 miles (34 km) long by 4 miles (6 km) wide. The two main towns of Vieques are Isabael Segunda (sometimes written "Isabel II"), the administrative center located on the northern side of the island, and Esperanza, located on the southern side. At peak, the population of Vieques is around 10,000.

The island's name is a Spanish spelling of a Native American word said to mean "small island". It also has the nickname *Isla Nena*, usually translated from the Spanish as "Little Girl Island", as a reference to

---

1. Even terrorists have websites. Most of the information on *los macheteros* came from http://www.latinamericanstudies.org/epb-macheteros.htm although some was from Manuel, and other bits came from various late-night trawls across cyberspace.
2. "Cheese" first told me about Vieques, but a damning indictment of WW2 encroachment there can be found at: http://www.sscnet.ucla.edu/soc/faculty/ayala/vieques/Papers/06ayalacentro.pdf and supportive information can be found at U.S Government archives http://www.atsdr.cdc.gov/sites/vieques/vieques.html

Island of Paradise

its being perceived as Puerto Rico's little sister island. During the colonial period the British name was "Crab Island", and in idle moments since that momentus winter meeting, I have wondered if this was the inspiration for the Crab Island in Arthur Ransome's children's novel *Peter Duck* (1930).

During the 1940s the island changed forever when the U.S. Navy purchased about $^4/_5$ of the available land on the island by compensated land expropriation. Later, the U.S. military started using them as target practice for bombs and missiles, and for other ground practices - like the USMC beach and helicopter infiltration exercises.

There have been some allegations that these practices are the cause of Vieques high cancer rate. The *World Socialist Web Site* (WSWS), published by the International Committee of the Fourth International (ICFI) – a vociferous opponent of U.S. military and foreign policy - reports that *"over a third of the island's population of 9,000 are now suffering from a range of cancers and other serious illnesses"*.

According to Manuel that was only the beginning. He told me that just after WW2, two covert American research projects were set-up. One on the island, and one in a little village by a lagoon on the mainland coast. He claimed that both of these projects were carrying out massively illegal research on human beings.

Nick Redfern's book *Bodysnatchers in the Desert* had only just been published, and had not - at the time of my long, soul-searching conversation with Manuel - received any great publicity. I had been aware of his research for well over a year. I still remember my shock when he showed me pictures, which he had found in the National Archives in Maryland, which appear to show someone with progeria having suffered terrible burns. If Manuel was telling the truth, it seemed more than possible that the post-war American experimentation, which followed on from the disgusting experiments by the Japanese during WW2 had not been confined to a few high altitude balloon exercises in New Mexico. Could this be an explanation for the strange village which both Rueben and "Cheese" had told me about? A village inhabited by deformed people with progeria-like symptoms. And if other - possibly more arcane - experiments had taken place, and may well still be taking place in the area, could the presence of some sort of research facility explain the strange experience that happened to Carola and her friends?

There was one thing that was puzzling me. Why was Manuel telling me all this? After all, our relationship was based upon a 20-minute conversation at JFK airport six months before, and a cursory exchange of eMails.

His explanation was both flattering, and frightening. The day that I had first met him, he was in the throes of leaving Puerto Rico for good. Over the years, his climb through the ranks of the National Guard had continued, and by 2004 he had achieved a respectable mid-ranking position. However, his father still had high ranking intelligence contacts, and one of his friends had warned him that his only son was in imminent danger of being arrested on terrorism charges. His father confronted him, and Manuel realised that there was no point in denying anything. He also realised - much to his surprise - that he and his father had more in common than he had thought. When he retired from the military, his father had become involved with local conservation projects, and had grown to feel an incredible guilt for his involvement in the Rainbow Herbicide projects.

Although, unlike Manuel, he would never have broken his conditioning to the extent of joining *los macheteros*, he was sympathetic enough to pull enough strings to allow his son to escape. When I blundered upon them at JFK, they were saying goodbye - perhaps forever. He had given Manuel my eMail address, and when Manuel eventually arrived in London, after a few months spent in Scandinavia, he had looked me up on the Internet, originally planning merely to hook up with one of the few people in England whom he knew was interested in Puerto Rican conservation issues.

However, even a cursory investigation of me told him two important things that he had not already known. Firstly, that I was one of the world's experts on *el chupacabra*, and furthermore, that I was one of the few cryptozoologists who took the subject seriously, and did not try to explain it with mumbo-jumbo, but tried to place it firmly within a zoological framework.

But he also knew that I had a reasonably deep knowledge of UFOs, although I am not particularly interested in them, and have usually shown public disdain for the extra-terrestrial hypothesis.

Secondly, that I had always been open about my anarchist principles. That I was interested in radical politics, and - after reading my autobiography - he figured that I was unlikely to be offended by being contacted by somebody from what appeared, to the outside world at least, to be a terrorist organisation.

He was tired of being a fugitive, he said. *"I am* machetero, *and I am a warrior. A warrior does not hide like a refugee in a foreign land. A warrior goes home to fight for what he believes in".*

He had made the decision to return to the Island of Paradise, to resume the armed struggle against the American oppressors. However, he was pragmatic. There was every chance that he would be arrested, or even killed. He wanted to tell what he had found out to somebody that - he hoped - would believe him.

But what has the chupacabra got to do with this? I asked. Was he saying, like the Las Vegas newspaper had claimed that I had said, that the chupacabra was actually one of the deformed half-people from the mysterious lagoon? Was he saying that the chupacabra was actually the result of some demonic biological experiment? Or could he even be suggesting that the goatsucker was somehow linked to the El Yunque UFO crash? After all, I had never mentioned UFOs to him. It had been him, not me, who had introduced the subject into the conversation.

Suddenly, he became angry. The gentle saint from an El Greco painting had been replaced - in an instant - by a fiery warrior; and furthermore, one with a particularly nasty temper.

"Do you think I am an ignorant fucking peasant, to be taken in by such primitive shit?" he almost screamed at me. *"Do you believe all those American lies that we are all gullible fools, fit only for manual labour and to be subjects of their Capitalist empire?"* and he spat, loudly and vehemently on the floor of the pub.

Suddenly, two large bouncers descended upon us, and none too politely asked us to leave. As we left the pub unceremoniously, Manuel suddenly calmed down, turned to me, kissed me on the cheek, and apologised. He asked me if I was hungry. I nodded, and we made our way to the *Yo Sushi* stall, next to the statue of Paddington Bear. Over raw fish and cold beer he told me the rest of the story.

Although, he said, that he could never forgive the American filth for the way that they had treated his countrymen, biological experimentation, and environmental degradation were not the worst of their crimes. *"Do you know why the pigs in Washington oppose giving us either statehood or independence?"* he asked. And without a waiting to give me a chance to reply, he continued. *"It is because they want to keep my people at a peasant mentality",* he said. *"Only then can they hope to hide the truth of what they have done from the rest of the world".*

Managing to get a word in edgeways, I told him what I had discovered about how the American army had utilised the Phillippino myth of the *aswang* during the 1913 war. I also told him about my belief that certain well-known UFO incidents, most notably the Rendlesham Forest incident of 1980, were nothing more than covert psi-operations on behalf of the American military. Independently of Manuel, I had come to the conclusion that fostering a fear of the paranormal had been a recognised part of American military strategy for the best part of a century.

He nodded agreement.

"There are no aliens. There are no grotesque blood-sucking monsters. But so long as my people believe in such things, then they can never be free. If they believe in - how do you say? - the bogeyman, then they will never go searching for the real truth. I have told you what I believe. You can, if you wish, tell the whole story. Indeed, I hope you will.

But you'll never hear from me again. You must go home, and I must go and fight a war".

We embraced, turned our backs on each other, and walked in opposite directions. I never have heard from him again, although I sometimes worry that I never fulfilled my part of the bargain. I started writing this book early in 2005, but events in my own life took over. My father was taken ill, and I left my home in Exeter to look after him until he died. I then inherited the old family home, and with all the complications inherent in moving house, nursing my father, and then dealing with wrapping up his estate, this book - which should have been finished during the summer of 2005 - wasn't finally completed until three years later. In the meantime *machetero* leader Filiberto Ojeda Ríos was ambushed and killed in the autumn of 2005. I have a sneaking suspicion that the *machetero* leadership was hoping that my book would be published on time. I have often wondered whether Filiberto Ojeda Ríos was actually planning a popular uprising when he was killed by the establishment. If the book containing Manuel's revelations *had* come out on time, would the *macheteros* have acted as an unofficial press office for me, and publicised it far more than I would otherwise have been able to, using me as an unwitting pawn in their revolution?

We will never know.

Back in 2004, I found that I had missed the last train back to Devon, and I spent - not for the first time in my life - an uncomfortable night sleeping rough on the station before catching the milk train back to Devon to spend the only Christmas I would ever spend with a young family, who were at least temporarily, my own.

Chapter Nine
Genesis in the island of paradise

My mother introduced me to the works of Rudyard Kipling when I was still a toddler. She started off reading me the *Just So Stories* when I was about four years old, and we soon progressed to the *Jungle Book*. I fell in love with his prose, and soon came to agree with my mother that his poetry - which always had the meter: 'tum te tum te tum te tum' - were unsurpassable.

By the time I was six and a half I could read it for myself, and at an age when my peers were struggling with godawful books explaining that Janet had a red ball, and that Spot the dog could run, I was immersing myself deep within Kipling's glorious usage of the English language.

I was particularly fond - and still am - of the poem *The Law of the Jungle* from the *Jungle Book*:

> Now this is the Law of the Jungle
> as old and as true as the sky;
> And the Wolf that shall keep it may prosper,
> but the Wolf that shall break it must die.
> As the creeper that girdles the tree-trunk
> the Law runneth forward and back
> For the strength of the Pack is the Wolf,
> and the strength of the Wolf is the Pack.

This was heady stuff for a seven-year-old, but it introduced me to one of the concepts by which I have lived my life ever since: that there are inexorable laws of the universe, with moral values which greatly surpass anything from the law of man.

Another poem, *Mowgli's song against people* from the story 'Letting in the jungle' also had a great effect on me.

> I have untied against you the club-footed vines
> I have sent in the Jungle to swamp out your lines!

> The trees-the trees are on you!
> The house-beams shall fall;
> And the Karela, the bitter Karela, Shall cover you all!

The story tells how Mowgli has been driven out of the human village for witchcraft, and the superstitious villagers are preparing to kill his adopted parents Messua and her (unnamed) husband. Mowgli rescues them and then prepares to take revenge. The karela (*Momordica charantia*) is a tropical and subtropical vine of the family Cucurbitaceae, widely grown for edible fruit, which is among the most bitter of all vegetables.

Most seven-year-olds believe that the world is permanent; that what is, always was, and always will be. Kipling taught me differently.

Another lesson from this remarkable book came in 1967 when Walt Disney produced a full-length animated feature film "based" on the *Jungle Book*. When I heard that it was going to be a film, I was so excited. At last I would see the visions from my head on the big screen. Like my school friends I went to see it at the cinema, but unlike them, I was appalled. Where were the characters I loved so well? Where was the deep, spiritual poetry? Why were the wonderful poems replaced by those stupid songs? And why - when all the other children were enjoying themselves, and singing along to the trite melodies - did I feel like bursting into tears?

I learned my third valuable lesson that day. That crass commercial concerns will override anything of genuine substance. That most people in the universe have absolutely no taste. And that corporate multinationals, like the Disney Corporation, are the enemy.

Nothing has happened in the intervening 42 years to change my mind.

So I went back to the books, and to my great joy I found that there were echoes of Kipling all around me. I don't know whether if I had spent my pre-teen years in England rather than in Hong Kong, whether this book would have had such an enormous effect on me. But in Hong Kong - or so I discovered, once I picked up my mother's copy of a book by Geoffrey Herklots called *The Hong Kong Countryside* - I was surrounded by creatures straight out of the pages of the two volumes of the *Jungle Book*.

There were no wolves in Hong Kong, but most of the other animals in the books - or at least their close relatives - were to be found there, or had been found there within living memory. Herklots wrote about the wild red dog, or dhole, which appears so frighteningly in the story *Red Dog*. Although he never saw one for himself, he wrote that: *"They are said to be reddish in colour and looking in the distance like Alsatian dogs, but with shorter legs"*.

Rikki Tikki Tavy was a great Indian mongoose. The species was absent from Hong Kong, but two closely related mongooses are found there to this day.

Even Shere Khan, the most fearsome of villains, could occasionally be seen. Although the last positive record of a tiger in the colony was in 1948, there were reports all the way through the 1960s and early 1970s that a lone specimen had strayed in from the Chinese mainland. The *Bandar Log* or monkey kingdom were represented by two species of macaque, and even the minor characters like the fruit bats and Ikki the porcupine will well represented.

I have always been particularly fond of a passage that describes how the small boys who herded the oxen in Kipling's India spent their days:

Island of Paradise

"Then they sleep and wake and sleep again, and weave little baskets of dried grass and put grasshoppers in them; or catch two praying mantises and make them fight; or string a necklace of red and black jungle nuts; or watch a lizard basking on a rock, or a snake hunting a frog near the wallows. Then they sing long, long songs with odd native quavers at the end of them, and the day seems longer than most people's whole lives, and perhaps they make a mud castle with mud figures of men and horses and buffaloes, and put reeds into the men's hands, and pretend that they are kings and the figures are their armies, or that they are gods to be worshiped. Then evening comes and the children call, and the buffaloes lumber up out of the sticky mud with noises like gunshots going off one after the other, and they all string across the gray plain back to the twinkling village lights."

Because in the halcyon days, which were the extinction burst of the greatest empire that the world has ever known, this is pretty well how the ex-pat children with whom I went to school spend their leisure time as well.

I am probably a representative of the last generation of English children who were lucky enough to live out the childhood in what would today be described as the 'Kipling Universe'.

In 1971 my family returned to England, and it was my private sorrow - one that I could never share with anybody else - that I had left the world of Kipling behind. But it wasn't so.

In 1878 Kipling was sent to a boarding school - the United Services College at Westward Ho! in North Devon. 93 years later I was sent to school in Bideford four miles down the road, and once again I found myself following in the master's footsteps.

In 1899 Kipling published a book of stories about his schooldays entitled *Stalky & Co*. In glorious, though slightly more hard-hitting prose than the *Jungle Book* he described the exploits of three schoolboys in the 1870s. A century later not much would change:

"...they flung themselves down on the short, springy turf between the drone of the sea below and the light summer wind among the inland trees. They were looking into a combe half full of old, high furze in gay bloom that ran up to a fringe of brambles and a dense wood of mixed timber and hollies. It was as though one-half the combe were filled with golden fire to the cliff's edge. The side nearest to them was open grass"

The English countryside was still beautiful, and the combes were still filled with golden fire to the edge of the cliffs. Something else hadn't changed. Most schoolboys were obsessed by sport, and those like me - and Kipling and his pals a century before - were treated with suspicion.

""Mad! Quite mad!" said Stalky to the visitors, as one exhibiting strange beasts. "Beetle reads an ass called Brownin', and McTurk reads an ass called Ruskin; and..."

"Ruskin isn't an ass," said McTurk. "He's almost as good as the Opium Eater. He says 'we're children of noble races trained by surrounding art.' That means me, and the way I decorated the study when you two badgers would have stuck up brackets and Christmas cards. Child of a noble race, trained by surrounding art, stop reading, or I'll shove a pilchard down your neck!" "

I was a 'child of a noble race'. I wanted to be part of a group of anarchic friends who appreciated great literature and high art as much as they appreciated behaving badly, and causing trouble. In that respect, like most others, my school days were a bitter disappointment. But I left school, got on with my life, and eventually - to a certain extent anyway - achieved my goals. But I never left Kipling behind.

After a long, cold night on Paddington Station I arrived back in Devon at lunchtime. Relations with my girlfriend were as bad as ever, but I threw myself wholeheartedly into making sure that the four children had a decent Christmas, because I knew it was highly unlikely that I would ever be there with them for another one.

Christmas was as horrible as I had expected it to be, and a few days after, the bombshell that delivered the final coup de grace to my relationship was dropped. The bombshell arrived in the form of a 60 old ex-convict who happened to be my girlfriend's estranged father. For reasons that it is not appropriate to explain, she had never met him before, but she welcomed him into our home with open arms. I disliked him intensely, and I'm sure the feeling was reciprocated.

Whilst my new father in common law held forth loudly at the dinner table about his experiences working in a slaughter house, telling the four children in graphic detail how he had been employed to cut the throats of pigs, I retreated back into my own inner marshland, taking solace in the comfort of the company of two of my closest friends; Jack Daniels and Rudyard Kipling. It was one night when I was reading the `Rikki Tikki Tavy` to the youngest child that I had an epiphany.

One passage from the second Jungle Book sparked it off:

"It began when the winter Rains failed almost entirely, and Ikki, the Porcupine, meeting Mowgli in a bamboo-thicket, told him that the wild yams were drying up. Now everybody knows that Ikki is ridiculously fastidious in his choice of food, and will eat nothing but the very best and ripest. So Mowgli laughed and said, "What is that to me?" "

But it was on, what turned out to be my last night with the children, that the penny dropped. I was reading *Rikki Tikki Tavy* aloud when I suddenly realised something very important. The previous summer, whilst I had been at the farm in the grassland plateau above Canóvanas, and had been told about the peacocks that had been killed by something biting a triangular shaped hole in the skull, I knew that - somewhere in the murkier parts of my subconscious - I knew what had killed them. Now - thanks to Kipling - I was sure.

The killer was a mongoose. And on that last night in Puerto Rico, in the nastiest suberb of Canóvanas, I had caught a brief glimpse of an Indian mongoose.

Although Herklots had written that the crab-eating mongoose *(Herpestes urva)* was a native to Hong Kong, it was last seen in about 1950, and was even then considered rare. An apparently healthy population was discovered by chance in 1988. But then - much to everybody's surprise - in 1990, a second species, the Javan mongoose (*Herpestes javanicus*), was discovered, in apparently larger numbers than the first species.

Over the years I've been collecting information on Hong Kong natural history for a book, which one of these days I'll get around to writing, and so I have a large archive - possibly the largest archive in the UK - on materials appertaining to the animals of Hong Kong. Because of the fortean aspects to their discovery, apparent extinction, and spectacular rediscovery, I have a particular interest in the two mongoose species. As I finished reading the story to the little boy, whom I knew that I would be deserting, much against my will, before too much time had passed, I was a psychologically kicking myself for not having realised the identity of the killer of Canóvanas.

I also now knew what had attacked the banana trees. Once again, Kipling had come to my rescue, and once again my life was following weirdly in a track defined by his precepts.

I kissed the little boy good night, and went downstairs. A friend of mine - an Irishman, who - if he had not been fair-haired - would have ideally fitted Kipling's description of M'Turk as "a dark and scowling

Indian mongoose (*Herpestes edwardsii*)

Celt" had come to visit, and he bought a bottle of whisky with him. He was - coincidentally - one of the ex-terrorists to whom I referred in the previous chapter, but he is a very dear friend of mine; a wise, intelligent, and deeply spiritual man. Unfortunately, when we meet up (which is sadly, not very often) we often end up getting drunk and singing Irish rebel songs.

Although my days of drink and drug abuse are largely behind me, I do - sadly - have a tendency to drink when I am unhappy, and that night - despite the welcome, and totally unexpected arrival of my close friend - I was very unhappy. So the two of us got very drunk.

We drank, and sang, and toasted to the memory of Fenian heroes long gone. The next morning I woke up with a terrible hangover, and my girlfriend threw me out! Apparently her father had disapproved of my behaviour, and she decided that whereas a convicted bank robber with an intimate knowledge of how to cosh nightwatchmen, and rip out the windpipes of stunned, though still living farm animals, was an appropriate role model for her children, whereas a fat hippy who knew all words to the *Boys of the Old Brigade* and the *Broad Black Brimmer* was not.

She wouldn't even give me a lift into town, so I telephoned for a taxi, grabbed as much of my stuff as I could carry, and left forever. I never saw any of them again.

Later that morning, still nursing a magnificent hangover, I waited outside the supermarket in the small market town for Graham, who - as always - responded magnificently to a crisis, and took me back to my little house in Exeter.

Within days I realised quite what an appalling mistake I had made. I had been so keen to throw myself into fatherhood, that I had neglected most CFZ activities for nearly six months. What was worse, I had spent nearly all my savings on my new family. Both my personal bank accounts, and the CFZ ones were empty. I had enough money to last a couple of weeks, but in very real terms I was facing financial ruin.

The boys of the CFZ have many faults: they can be shiftless, lazy, drunk, truculent, and unmotivated. But whenever there is a crisis, they are bloody magnificent! Slowly, over the next weeks and months, we clawed our way back from the brink of disaster, and emerged from the crisis stronger, bigger, and better than ever.

Now I no longer had the day-to-day trials and tribulations of a young family, I had time to reconsider my theories about the mystery animals of the Island of Paradise.

I missed the four children dreadfully, and - in a way - over three years later, I still do. But I didn't miss the continual psychodrama that was inherent in living with my ex girlfriend. I didn't miss the moral dilemma in which I continually found myself when dealing with her peculiar beliefs on religion. And I positively embraced the chance to throw myself wholeheartedly back into my research.

A week or so after my return to Exeter, Darren Naish got back in touch with me. Whereas he was convinced that there was an obscure species of porcupine known from one subfossil record from somewhere in the Lesser Antilles, he hadn't been able to track down the reference. However, he was one step closer to finding a West Indian porcupine. A little known species called the pallid hairy dwarf porcupine had been described in 1848. The IUCN red data list notes:

"MacPhee and Flemming (1999) disagree with the inclusion of *S. pallidus* on the list of extinctions until the taxonomic uncertainties surrounding the species have been resolved. The species is known only from two immature specimens which probably came from somewhere in the West Indies (range states unknown). The matter has been referred to the relevant Specialist Group for a decision."

Another acquaintance of mine was Ken Smith, a student of zoology at Universtiy in the Midwest. I was talking to him late one night on the telephone about my problem, and he told me a little bit more about the mysterious Antillean porcupine. It had indeed been discovered from one subfossil specimen found somewhere in the Lesser Antilles. It had been found in the early part of a twentieth-century by a German palaeontologist, and the remains had been taken to Dresden Museum in Germany for further study. However, as we all know, Dresden was bombed flat by Allied bombers during WW2, and as far as he knew, the specimen had been lost. He promised that he would find out as much as he could for me, and put down the phone. Sadly, he died in a car crash a few months later, whilst I was caring for my dying father, and the trail went cold.

By this time I was convinced that the mysterious vampiric attacks on livestock, and the sightings of the strange spiky beast, were two totally different phenomena. I am certain that mongooses did the vast majority of the killings of domestic livestock. The clincher for me was that - according to Ismael and my other friends on the island - there had not been a verifiable vampiric attack since 1998. This coincided with other things I had been told about an explosion in the numbers of rats on the island in the mid-1990s. Unlike rodents, carnivores increase in population numbers exponentially depending on the amount of available food. This is one of the reasons that fox hunting is pointless as a method of keeping down the numbers of foxes in a specific area. If a hunt kills - say - six foxes in a season, then the fox population will breed to replace those six foxes. The size of the population will only increase if there is an increase in the volume of available food animals.

Other small animals work in the same way. If there were, indeed, an explosion in the rat population of of Puerto Rico, for whatever reason, then it would seem highly likely that the mongoose population of the island would increase shortly afterwards. If - for whatever reason - the rat population decreased again, then - for at least one generation of mongoosedom - the viverrid population of the island would have to have turn to new and unusual prey species. The fact that the killings seem to have stopped after a period of only three or four years, would seem to indicate that the mongooses did not really adapt to this new way of life, and that - a generation or so later - the mongoose population of the Island of Paradise returned to its normal levels.

And, after having seen the peculiar, tiny goats of the Canóvanas plateau for myself, I was convinced that a determined mongoose could easily be a match for one of them.

The following passage, concerning a rare Sri Lankan mouse deer having a face off with a brown mongoose, appeared on the Wildlife Direct website in march 2008, just as I was putting this book to bed. I quote it in full, because it provides conclusive proof that a mongoose *will* attack an ungulate – even one several times its size:

A mountain mouse-deer was seen under quite dramatic circumstances in February 2008 by wildlife photographer and specialist Gehan de Silva Wijeyeratne, and naturalist Nadeera Weerasinghe. While providing a training session on butterflies and dragonflies for the staff of the Horton Plains National Park, an animal came running and jumped into the pond and swam towards them. It was identified as a mountain mouse-deer, being pursued by a brown mongoose, about a third of its size.

The mouse-deer swam back to the far shore and faced off with the mongoose. The mongoose did not enter the water but at times approached within five to six feet of the mouse-deer which responded by flaring its throat and showing the white on its throat.

After fifteen minutes the mongoose seemed to tire of the chase and left. The mouse-deer left but returned soon with the mongoose in pursuit and once again dived into the pond. Forty-five minutes later the duo left and Gehan de Silva Wijeyeratne and Nadeera Weerasinghe informed the park warden. Around 5 pm the mouse-deer was seen again by the park warden and his staff. Later around 6pm it was taken in for safe custody, and offered no resistance. It had a small gash near the ear and was in an exhausted state.

I believe that my own observations of Puerto Rican goats, combined with this testimony from Sri Lanka is the last nail in the coffin of the theory that Puerto Rico holds a hitherto undiscovered vampiric animal. The universally held assumption that the vampiric marauder would have to be able to kill something the size of a British sheep, is also a complete fallacy, and the last argument against the idea that a mongoose could be responsible, went out the window.

However, for a few years, farmers were shocked to see their livestock - and remember, this is on an island where there are no natural predators - being killed in what they interpreted as a grotesque and demonic fashion.

Exit the vampiric killings. As far as I'm concerned, we can now take them out of the equation. There are - I have to admit - a number of unexplained killings of domestic livestock that cannot be explained using a theoretical model whereby mongooses are the culprits. However, the global cattle mutilation mystery is at the present at least unsolved, and I do not believe that it falls within the remit of this book to discuss it further. Other killings which have been described by researchers such as Scott Corrales seem to me to be human in origin, and are attributable either to cultists, devotees of one of the Santeria religions, or purely down to the fact that as the t-shirt worn by the farm labourer that we met high on the Canóvanas plateau proclaimed: People=shit!

But what about the resemblance of the chupacabra to The Devil? My discovery that the Latino depictions of the demonic host made them almost rodentine in appearance, was merely supportive evidence, that the spikey creatures that had been reported over the years were undoubtedly porcupines, or - at least - a close relation.

But what about the porcupines? As I have described earlier in this book, the porcupine family is divided firmly into two parts, and the New World porcupines are almost exclusively arboreal. If the animal of the Canóvanas grasslands is a porcupine, then it would appear on the face of it to be far more allied - morphologically speaking, at least - to the Old World porcupines. Is this an insurmountable problem? Or can we find an explanation?

Until we actually obtain a specimen, we will not be able to provide a definitive answer. However there are a number of hypotheses that warrant examination.

Firstly, as we have shown elsewhere in this book, parallel evolution is a remarkable process. As I discovered, when just a lad in Hong Kong, porcupines do have a penchant for attacking the trunks of plantain trees. And - as we have also seen - Puerto Rico is knee-deep with plantain trees. Could a New World porcupine have evolved to fill an ecological niche, familiar to its distant cousins in Africa and Asia? It is not beyond the bounds of possibility.

Secondly, as can be seen in this extract from a 1919 paper on the extinct mammals of Puerto Rico, which is printed - in its entirety - as Appendix One to this present volume, several of the larger rodents of the Lesser Antilles, which are presumed extinct, were closely related to porcupines:

"The rodents all belong to the Hystricomorph or porcupine section of this order, which has its headquarters in South America.

Island of Paradise

The largest of them, Elasmodontomys, is closely related to Amblyrhiza, the gigantic extinct rodent of the island of Anguilla. Another smaller genus, Heptaxodon, also belongs in this neighborhood, but is less closely related. These are put into the family Chinchillidæ, but their nearest affinities are with the extinct Megamys and Tetrastylus of the Pliocene of Argentina."

We don't actually know what any of these species actually looked like in life. Could one of them, or a hitherto undiscovered close relative, have appeared - on a cursory level at least - to resemble one of the Old World porcupines? And could one of them have survived to the present day?

Thirdly, we have the speculation that Puerto Rico was a Microcontinent; not breaking off from mainland South America, but drifting - in its entirety - across the Atlantic Ocean from Africa. It would be ludicrous to suggest that it brought with it a fully formed species of Old World porcupine, but it might have brought with it something of an Old World eco-system that would provide enough environmental triggers to push the evolutionary progress of an Antillean porcupine in a hitherto unsuspected direction.

Fourthly, we might actually have an Old World porcupine living on the island. Porcupines are, after all, highly adaptable beasts. At least two species have - for a time at least - lived wild, and reasonably successfully, in Britain. Writing in 1975, Sir Christopher Lever described how a population of Himalayan porcupines escaped from a badly run wildlife park near Okehampton in mid-Devon in 1969, and set up a successful colony near Dolton. Other reports of this species have been made across the south of England, and Lever also notes a sighting of the African crested porcupine somewhere on Salisbury Plain.

As we have seen, many alien creatures - and here, probably needlessly, I want to stress that, despite the fact that a proportion of this book is discussing the UFO phenomenon on the island, I am using the a word in its zoological context, to describe animals that have been imported by accident or design from elsewhere - have become established on the Island of Paradise. Norka told us about giant snakes - presumably pythons or boas - which have become denizens of parts of the forest. I have already described how monkeys from Asia became established on one part of the island, and as Old World porcupines are popular as zoo animals, it is certainly not beyond the bounds of possibility that some have escaped, or been released from captivity to find a new home in the high mountain grasslands of the Canóvanas plateau.

But there is yet *another* possibility. Puerto Rico was one of the main stopping-off points for slaves imported from Africa. Several porcupine species are well known food items in West Africa, and it is not beyond the bounds of possibility that some live specimens may have been taken on board the slave ships as potential meals, and then - accidentally, or on purpose - released upon their arrival on the island.

I believe that the sightings of what I once described as 'Sonic the Hedgehog on acid', are of a porcupine, or porcupine-like species, which is probably unknown to science. And I believe that the reason that none of the expeditions that have gone in search of the chupacabra have succeeded, is because they have been looking in the wrong place for the wrong thing. They have been looking for a nocturnal vampire, when they should have been looking for a crepuscular rodent.

It is, I think, important that the further away one gets from the Canóvanas plateau, the more outlandish the reports become. I believe that I have found a very real explanation for the core chupacabra sightings, and that most of the others are folkloric, or - I grudgingly have to admit - paranormal in nature.

However, I also freely admit, that this theory does nothing to explain the sightings of an owlman-type creature as reported by Norka or Nick Redfern's interviewee. But, having fallen flat on my face in the past whilst attempting to do so, I long ago gave up trying to formulate unified field theories of everything. And I know, to my personal cost, that weird shit does, indeed, happen. The reports I received from the old man in Canóvanas, and from Benny in Las Vegas were too similar to my own experience at Bolam Lake for comfort, and I was sure that they were telling the truth. And there were reports of owlman-like creatures from all over the globe, and - disturbingly - since the publication of my most famous book, people send me more on a regular basis. There are a number of other mysteries on Puerto Rico beside that of the chupacabra, and beyond the scope of this book.

So I will ignore most of them, because as I approach my half-century I am becoming increasingly aware that some issues are unknowable, and I have a sneaking suspicion that - at least in my lifetime - we will never know the truth behind the owlman and his kin.

However, one mystery, which I do find eternally fascinating, is that *something* appears to be causing accelerated evolution at a ridiculous rate amongst the inhabitants of the Canóvanas River.

I have isolated four species that that seem to have become affected.

- Firstly, the gambusia. This is a well-known laboratory animal, and as I described earlier in the book, has been kept in captivity for well over a century. Its biology is well-known, as are its colour morphs. I believe that I am the first researcher to suggest that - for some reason, presently unknown - a population of these charming little fish has evolved with a peculiar snub nose. They are still recognisably gambusia, but they are unlike any gambusia I, or indeed any other aquarist, has ever seen.

- Secondly, the cave snails. Once again, they are recognisably members of the same root species as the ubiquitous forest snails. However, they are recognisably different. And, unless my admittedly meagre knowledge of mollusc biology is seriously awry, they would have had to make some serious biochemical and endocrinal adjustments in order to adapt to a troglodytic lifestyle, and a diet of guano.

- Thirdly, the freshwater mantis shrimps. As I have shown, the morphological adaptation required to adapt a shrimp species to be able to hunt like a mantis is a common one. But until now, it is one that is not known to have occurred in fresh water.

- Fourthly, my new lamprey species. The more that I research into lamprey biology, the more I am becoming disturbed by the implications of my original claims to have found the first tropical species of lamprey. The known evolutionary biology of these fish would seem to suggest that they can only be found in warm water. What I am beginning to wonder is whether the little fish that I caught back in 1998 was in fact an eel - but an eel with a morphological adaptation that is known from catfish species, and some loaches, but never before from a member of the Anguilla family; paired pectoral fins which have become adapted into a sucker. Fish such as the hillstream loaches of South-east Asia exhibit this adaptation, which is necessary for life in fast flowing streams. Could the reason that the children of Canóvanas had never seen such a fish, be that the one I caught had been swept down from the mountain streams which are its natural habitat, and that my specimen was seriously out of place.

Occasionally fish *do* stray from their natural habitat. I always remember when, as a child, I was fishing at Venn River - the tiny stream, a few miles from my home, which I described in an earlier chapter. Over the years I only caught four species of fish over a three-mile stretch of water. These were three described earlier: bullheads *(Cottus gobio)*, stone loach *(Nemacheilus barbatulus)*, and brook lamprey *(Lampetra planeri)*, plus occasional brown trout *(Salmo trutta)*. But one day in 1975 I caught a large male minnow *(Phoxinus phoxinus)* in breeding colours; a species I never saw in that river before or since, and it was a river that I studied in great depth for many years. Where it came from, I don't know. Could the same thing have happened that January day in Canóvanas ?

There is, however, one big difference between the first two species on this list, and the second two. In the case of the first two, the root species is clear. What isn't clear, especially as far as the gambusia are concerned, is the evolutionary advantage conferred upon these little fish by having a snub-nose.

In recent years, much of my work with the CFZ has involved the European eel *(Anguilla anguilla)*. We have come to believe that most of the so-called lake monster sightings across the northern hemisphere are gigantic eels, and following a spate of sightings of what appear to be enormous eels in the Lake District, we have carried out two trips to date searching for them.

On our first trip to the Lake District in October 2006, we didn't see a single eel. On our second trip, we did a little better. Richard saw two and Kevin (our diver) saw three or four. They were silver grey in colour, and well within the accepted size reference for this species. We were actually quite disappointed, because although Kevin and Richard had snorkelled pretty well solidly for two days, and Kevin did several hours of diving, not only did we not manage to come back with any decent underwater photographs, but the fish population as a whole seemed to be remarkably sparse.

Some weeks after returning, I received a coldly polite telephone call from a gentleman working in the Environment Agency. *'Did we know'*, he asked us, *'that we had been breaking the law by setting eel traps without a licence?'* *'No'*, I said, *'and even in my most recklessly anti-establishment moods I would not have been stupid enough to knowingly break the law, and then release photographs of me doing so to the gentlemen of Her Majesty's press'.* This broke the ice, and he gave a slightly nervous, civil-servantish laugh. We talked for a while about what we had hoped to achieve at Coniston, and he told me that, in recent years, there has been an enormous decline in the eel population.

This got me thinking.

More out curiosity than out of any other reason, some weeks before we went to Coniston I downloaded the Project Gutenberg version of Isaak Walton's seminal *The Compleat Angler*.

Walton was a consummate naturalist and his observational skills cannot be faulted. If we ignore, for the moment, his assertions about barnacle geese, goose barnacles, and over-wintering swallows, most of the rest of what he says about the lifestyle of the common eel is borne out perfectly by what we know today.

However, there were one or two items of interest. Any angler will tell you that the common eel *(A. anguilla)* has two main physical morphs. Although they are the same species, some eels develop pointy noses; these fish feed on worms and other invertebrates. However, other eels develop a blunt facial aspect, and these eels feed predominately on other fish. These two physical morphs have been known for well over a century, but it is interesting to find that Isaak Walton does not mention them.

But, he does mention several other morphs; one with a flat head much bigger than that of an ordinary eel, and one with reddish coloured fins. It is irresistibly tempting to deduce from this, that the two currently observed morphs of the common eel are relatively recent occurrences. Walton's other two morphs are not, as far as we are aware, found today.

Could it be that *Anguilla anguilla* is a very motile species? A species that produces new forms much more readily than would otherwise be assumed?

It is also interesting to note that whereas Walton remarks that eels do indeed go down to the sea, he claims that some give birth to living young in freshwater. The complicated life cycle of the European eel was only discovered in the 1920s, and the life cycle of the Japanese eel - a closely related species, *Anguilla japonica* - was only discovered within the last few years. According to accepted wisdom, the European eel mates, spawns and dies in the Sargasso Sea, located - amusingly enough - in the midst of the Bermuda Triangle. The Japanese eel, however, does exactly the same thing in the Marianas Trench near the Philippines.

It has been suggested by a number of our colleagues, that with the increasing number of land-locked stretches of water that some eels may well stay there for their entire life cycle and give birth to live young in freshwater. The comments by Walton, nearly 400 years ago, would seem to bear out this hypothesis.

For many years, it was thought that the European eel was unable to eat in saltwater. Relatively recently, it has been found that a significant proportion of the eel population do not seem to enter freshwater at all. This is again suggestive that our theory that eels are capable of changing both their form, and their lifestyle, very quickly, would appear to be correct.

On our way back down south, after three (mostly) fruitless days eel-fishing, we stopped at Blackpool Tower Aquarium. At the Big Cats Conference in Hull during March 2007, our old friend and compadre, Mark Martin told us that he had seen a particularly large eel - one that he estimated to be at least five feet in length in this particularly run down and unpleasant aquarium.

I don't like Blackpool. Whereas once it may well have been a magical place of delight, in many ways these days I feel that it is indicative of everything that is most wrong about British culture at the beginning of the twenty-first century.

An old friend of mine, who had grown up in Blackpool in the 1940s, took great exception to my venomous description of the place in my autobiography. This time, I went back with an open mind, hoping that I could prove him right. Admittedly, on my first visit, it had been both the Queen's Golden Jubilee weekend, and the height of the World Cup, and the behaviour of many of the people we saw walking down the streets was just as bad as one might expect. On this occasion, the behaviour of the populace was nowhere near as appalling, and I have to say that I can see how it had a certain faded charm.

But the drug addict panhandling outside the Blackpool Tower, the disabled beggar swearing sadly to himself under his breath, and the hard candy genitalia on sale to small children on a stall open to the street on the Golden Mile, however, only served to make me realise - as if any realisation was necessary - that Blackpool is hardly the place that I would wish to take my future grandchildren on holiday.

The Tower Aquarium itself was a complete shambles. None of the staff knew anything about the exhibits, and the main display tank was empty, after - allegedly - all the inhabitants had been killed after an unfortunate cock-up with the water chemistry. However, there were eels. And then some.

In one tank, there was an eel of about three foot in length, which - apparently - had been in captivity there for over thirty years. Here we have our first anomaly. The idea that eels die after 6 - 10 years has been well and truly exploded.

However, in a large saltwater tank there were two enormous European eels. Measuring them against my walking stick, pressed against the thick plate glass of the tank, they were considerable over 4 foot in length and the larger might even have been approaching five. Again, this is important. Both specimens are far larger than anyone would have any reasonable hope to expect. Granted, they are not as large as the fifteen - twenty foot eels reported by witnesses across the Lake District, but they are far bigger than they have any right to be.

These reports are only a work in progress. The big fish project is going to run and run and we are planning to be involved in research at the Lake District for at least the next five years. Even at this early stage, it seems certain that the biology of the European eel is not just poorly understood, but that it may have considerable surprises in store for us.

If we are correct, and *Anguilla anguilla* is indeed a motile species that can throw off new morphs very easily, then perhaps - as a result of the worldwide population crash throughout the species in the last few years - a new morph is occurring; one that is far larger than any eel currently known to mankind.

Bernard Heuvelmans once wrote that "there are lost worlds everywhere", but to find such a lost world in Blackpool Tower Aquarium beggars belief, even in this infinitely surrealchemical universe. Five-foot eels in a run-down tourist attraction, may not, as yet, provide the answer to the enduring mysteries of lake monsters across the northern hemisphere, but it is a damn good place to start!

In the case of the eel, the evolutionary advantages of these changes in their basic morphology are fairly obvious. But without serious further investigation of the tiny population of the snub-nosed gambusia, it is hard to see what evolutionary advantage could have provoked such a major mutation.

The case of the cave snails is much easier to argue. Simply, there was an entirely new food source to exploit, and exploit it they did. However, the internal biological changes would have to have been pretty massive, and this hitherto unknown adaptation is one that needs serious investigation.

The second two species are more problematical. In these two cases we do not know what the root species was, but as there are several species of eels known from Puerto Rico, and I have personally seen at least three species of normally shaped freshwater shrimp, one can take an educated guess.

If I am correct, therefore, there are at least four species of totally unrelated animal, which have undergone quite unexpected morphological changes. And furthermore, the one thing that they all have in common, is that they inhabit the same stretch of mountain stream, or - in the case of the cave snails - the subterranean area through which the mountain streams system flows. And - although this takes a somewhat of a leap of faith - I will not be at all surprised to find that the entire stream system flows from the tiny brook that wells out of the ground at what three different witnesses have told me was a UFO crash site.

Much to my surprise, although he raised one eyebrow and shot me what Richard Freeman calls 'a look if doubt and inquiry' at the concept that a UFO had anything to do with it, Darren Naish told me that my theory was a workable one; that some foreign substance which had somehow got into the eco-system could have provoked a range of apparently disparate mutations.

Sadly, science has a distressing tendency to mirror religion, and descend into fundamentalism. Because of the increasing attacks upon mainstream science by those members - usually of - the religious right wing, who espouse the - to me - frankly ludicrous doctrine of Young Earth Creationism, some scientists tend to adhere to the concept that there is no God but Darwin, and that Russell was his prophet. However, many scientists - some in the mainstream - now believe that the theory of evolution is more complicated than Darwin originally thought.

Punctuated equilibrium is a theory in evolutionary biology. It states that most sexually reproducing populations will show little change for most of their geological history, and that when phenotypic evolution does occur, it is localized in rare, rapid events of branching speciation (called cladogenesis). [1]

Punctuated equilibrium is commonly contrasted against the theory of phyletic gradualism, which states that evolution generally occurs uniformly and by the steady and gradual transformation of whole lineages (anagenesis). In this view, evolution is seen as generally smooth and continuous.

In 1972 paleontologists Niles Eldredge and Stephen Jay Gould published a landmark paper developing this idea. Their paper was built upon Ernst Mayr's theory of geographic speciation. I. Michael Lerner's theories of developmental and genetic homeostasis, as well as their own empirical research. Eldredge and Gould proposed that the degree of gradualism championed by Charles Darwin was virtually nonexistent in the fossil record, and that stasis dominates the history of most fossil species.

I am convinced this theory is broadly correct, however, the idea of at least four completely disparate species undergoing a bout of cladogenesis at the same time, does seem a little unlikely.

I have never made any secret of my religious or political beliefs, but I have usually kept them to myself. They have very little bearing upon my scientific work. I am a churchgoer and I believe in both a creator God - alpha - and the intercessionary God - omega. However, I see no contradiction between this belief, and my belief in the Big Bang theory, and in the broad precepts of evolution. The idea that the universe is only 6,000 years old is frankly ludicrous, and I even reject most of the concepts which people have come up with and grouped together under the broad heading of intelligent design.

According to Wikipedia (the free encyclopaedia):

"Intelligent design is the assertion that "certain features of the universe and of living things are best explained by an intelligent cause, not an undirected process such as natural selection." It is a modern form of the traditional teleological argument for the existence of God, modified to avoid specifying the nature or identity of the designer. Its primary proponents, all of whom are associated with the Discovery Institute, believe the designer to be the God of Christianity. Advocates of intelligent design claim it is a scientific theory, and seek to fundamentally redefine science to accept supernatural explanations."

I merely believe that although the Big Bang theory is completely correct, something - not someone, for it is anthropomorphic verging on blasphemy to create the Almighty in our image - had to (using words that will be familiar to anybody in Britain has ever bought fireworks) light the blue touchpaper. I believe - that as far as most of the claims of the proponents of intelligent design are concerned - that they made no sense. Why would the being that composed the rules of the universe constantly mess around with his creation?

However, my work in cryptozoology has led me to believe that new species are being created far more often than we think. My visit to Texas in November 2004 to investigate the blue dog of Elmendorf has led me to the conclusion that these creatures are a new species in the early stages of evolving from a known root species; the coyote. My conclusions as to the possible speciation of the European eel can be found earlier in this chapter. However, there is one particular syndrome of evolution which I believe has never been noted before, and is one for which cryptozoology can be thanked.

---

1. Everything I know about punctuated equilibrium came from conversations with three wise men: Richard Freeman, Chris Moise and Dr. Darren Naish. However Gould, S. J. (1992) "Punctuated equilibrium in fact and theory." In Albert Somit and Steven Peterson *The Dynamics of Evolution*. New York: Cornell University Press. pp. 54-84. Was very useful as well.

In the second half of the 20th Century three totally unrelated species of cats appeared to be undergoing a process of speciation.

- The onza of Mexico appears to be evolving from the puma.

- The Kellas cat of Scotland appears to be evolving from a complex introgressive hybrid of the Scottish wildcat and the domestic cat.

- And the king cheetah - a colour morph of the cheetah, has enough behavioural differences from its root species for many researchers to suggest that it is beginning the process of speciation.

The interesting thing about these three cats is that although they live on three different continents, and are only very distantly related, they are undergoing very similar morphological and behavioural changes at roughly the same time. All three are far more gracile than their root species, and all three appear to hunt in pairs - unlike their root species.

It is almost as if there is something deep within the genetic coding of cat species prompting them to speciate either at roughly the same time, or in response to the same environmental triggers.

This is not the time or place to speculate on what those environmental triggers might be, but in the case of the four species which I have identified from the hill streams of *El Yunque* I feel that we have enough evidence to do so.

Any one of a hundred environmental triggers might be responsible, but I would like to have examined just two.

It seems fairly certain that - using the purest sense of the term - an unidentified flying object crashed into the hillside in 1957. But what was it? For the purposes of this book, at least, we can ignore any assertions that it was a spacecraft from another galaxy. Firstly, there is no evidence whatsoever that such things exist, and in the absence of any such evidence it is pointless to speculate. Secondly, as I have shown in the previous chapter, there is a certain amount of evidence to support the suggestion that certain factions within United States military has been using the fear of the paranormal as a method of social, and military control for the best part of a century.

A very good friend of mine is called Larry Warren, and when he was a young man in the US Air Force, he was stationed at the air force base in Suffolk at the time of the alleged UFO crash in Rendlesham Forest. He believes that he saw a UFO land in a field deep in the woods. However, something that is less well publicised is that he has been having such experiences all his life. I have known Larry for over a decade, and when he first told me about his paranormal experiences, I believed that he was one of those lucky - or unlucky, depending on which way you look at it - people to whom weird shit happens. In my book the *Rising of the Moon* (1999) I included the theory that certain people, and indeed certain places, seem more susceptible to experiencing the phenomena that we broadly gather together under the umbrella of 'the paranormal'. I still believe that to be true, but in Larry's case I believe that the truth is far more sinister.

I think that since birth he has been one of the people earmarked by some covert group within the US government to receive a lifetime of psychological experimentation. I think that all his life he has been subjected to a series of artificial experiences designed to find out the effect of mass hallucinations and delusions upon the human psyche.

I finally came to this realisation at the 1996 Weird Weekend, when he, Nick Redfern, Mark North, my wife and I were sitting in my garden at midnight. Suddenly a huge triangular object flew out of the darkness, and over the field next to my house. All five of us saw it, and it was close enough for us to see a line of lights along the fuselage. It appeared to fly about 100 ft above ground level, and disappear into the night. Whereas we might have seen a UFO, I have a sneaking suspicion that if someone had been upon the main road at that moment, there would have been an unmarked van with a large satellite dish upon it, broadcasting a hologram into the air above my garden. I am of no defence significance whatsoever. Although Nick Redfern found that during 1996 and 1997 there was indeed a Special Branch investigation on me, it was because of my high profile espousing of the Irish Nationalist cause, which culminated in my appearing in a British Sunday newspaper wearing a Sinn Fein t-shirt. This was during the last days of a Conservative government which had lasted for 18 years, and which was getting increasingly paranoid. My Irish revolutionary obsession was partly boyish high spirits, and partly bipolar drunkenness. I was given a clean bill of health, and as far as I know, I have never been investigated since. However, Larry has been the subject of intense scrutiny by the powers that be for many years, and the fact that the only night we ever saw a UFO and fly over our garden, was the first night that he happened to be there, seems too much of a coincidence to ignore.

I tend to agree with Manuel. I think that the vast majority of the UFO activity which has been reported in Puerto Rico over the last 50 years is man-made, and that most of it has been deliberately engineered by the United States military in order to - as Manuel so succinctly put it - help keep the peasantry in a state of fear and superstition, so they never reach the stage of self determination which would allow them to discover, once and for all, quite how badly they had been treated over the years.

So if it wasn't an alien spacecraft, what was it? It would have to have been either a man made artifact, or a meteorite. If it was man-made, could it have contained large substances of one of the chemicals that have known to have been used in military experiments on the island since the 1940s? If so, would there have been enough chemicals on board to make any real difference? If it *was* a meteorite, could it have been comprised of a substance powerful enough to wreak genetic havoc upon the local ecosystem? Or could it have uncovered a buried deposit of something that probably should *not* have been uncovered?

Or are we barking up the wrong tree?

If you like to believe my findings then you will have already agreed with me that the chupacabra has nothing to do with UFOs, fortean phenomena, the paranormal, or whatever evolutionary abnormalities are going on in Puerto Rico. The chupacabra is - as I have always suspected - a purely zoological enigma, albeit a fiendishly complicated one.

Could the UFO crash be yet another red herring? And what about the strange village by the lagoon? Is it just a coincidence that - according to several of my contacts - this is where the Canóvanas River meets the sea. Could there really be a river that has a crashed UFO at its source, and a secret government research facility at its mouth? Isn't that *too* much to believe?

Is it not easier to accept that 30 or 40 years of intensive chemical testing of carcinogenic chemicals, and possibly even more destructive and vile substances has wreaked its own havoc upon the environment? My examinations of the leaf litter in El Yunque during 1998 showed that the area where I was at least was practically devoid of life.

Manuel certainly seemed to think so.

Other parts of the jungle are as rich in biodiversity as any other jungle I have been to across the globe. So

Island of Paradise

why are these areas so barren? Is it a coincidence that this specific area is adjacent to where the UFO allegedly crashed? Could the alleged UFO actually have been a controlled test of a missile containing some disgusting defoliant? There was certainly no plant growth there in 1998 - 41 years after the alleged crash.

My research into the eel population of the lakes of northern England suggests that some species become motile, and throw off new morphs when the root species is threatened. Could that have been the case here? It would seem unlikely, however, because the stream systems which houses three out of my four target species, are as zoologically diverse as any streams I have ever seen.

Or perhaps there is another answer.

500 years ago a tribal shaman entered into an uneasy friendship with the warrior from across the sea. He told him that the waters of the El Yunque rainforest were sacred, because they were the fount of all life; they were the waters from which all the animals first evolved.

Well I don't agree with Bishop Usher [1], and I don't believe Agüeybaná's theory either. But maybe it's not *all* wrong. Maybe there *was* once something special about some of the waters which flow from the steep black mountainsides of El Yunque. And maybe, just perhaps, there still is! After all, Puerto Rico *is* the Island of Paradise!

---

1 **James Ussher** (sometimes spelled **Usher**) (4 January 1581–21 March 1656) was Anglican Archbishop of Armagh and Primate of All Ireland between 1625–1656. He was a prolific scholar, who most famously published a chronology that purported to time and date Creation to the night preceding October 23, 4004 BC.

THE CENTRE FOR FORTEAN ZOOLOGY

So, what is the Centre for Fortean Zoology?

We are a non profit-making organisation founded in 1992 with the aim of being a clearing house for information, and coordinating research into mystery animals around the world. We also study out of place animals, rare and aberrant animal behaviour, and Zooform Phenomena; little-understood "things" that appear to be animals, but which are in fact nothing of the sort, and not even alive (at least in the way we understand the term).

Why should I join the Centre for Fortean Zoology?

Not only are we the biggest organisation of our type in the world, but - or so we like to think - we are the best. We are certainly the only truly global Cryptozoological research organisation, and we carry out our investigations using a strictly scientific set of guidelines. We are expanding all the time and looking to recruit new members to help us in our research into mysterious animals and strange creatures across the globe. Why should you join us? Because, if you are genuinely interested in trying to solve the last great mysteries of Mother Nature, there is nobody better than us with whom to do it.

What do I get if I join the Centre for Fortean Zoology?

For £12 a year, you get a four-issue subscription to our journal *Animals & Men*. Each issue contains 60 pages packed with news, articles, letters, research papers, field reports, and even a gossip column! The magazine is A5 in format with a full colour cover. You also have access to one of the world's largest collections of resource material dealing with cryptozoology and allied disciplines, and people from the CFZ membership regularly take part in fieldwork and expeditions around the world.

How is the Centre for Fortean Zoology organized?

The CFZ is managed by a three-man board of trustees, with a non-profit making trust registered with HM Government Stamp Office. The board of trustees is supported by a Permanent Directorate of full and part-time staff, and advised by a Consultancy Board of specialists - many of whom who are world-renowned experts in their particular field. We have regional representatives across the UK, the USA, and many other parts of the world, and are affiliated with other organisations whose aims and protocols mirror our own.

I am new to the subject, and although I am interested I have little practical knowledge. I don't want to feel out of my depth. What should I do?

Don't worry. We were *all* beginners once. You'll find that the people at the CFZ are friendly and approachable. We have a thriving forum on the website which is the hub of an ever-growing electronic community. You will soon find your feet. Many members of the CFZ Permanent Directorate started off as ordinary members, and now work full-time chasing monsters around the world.

I have an idea for a project which isn't on your website. What do I do?

Write to us, e-mail us, or telephone us. The list of future projects on the website is not exhaustive. If you have a good idea for an investigation, please tell us. We may well be able to help.

How do I go on an expedition?

We are always looking for volunteers to join us. If you see a project that interests you, do not hesitate to get in touch with us. Under certain circumstances we can help provide funding for your trip. If you look on the future projects section of the website, you can see some of the projects that we have pencilled in for the next few years.

In 2003 and 2004 we sent three-man expeditions to Sumatra looking for Orang-Pendek - a semi-legendary bipedal ape. The same three went to Mongolia in 2005. All three members started off merely subscribers to the CFZ magazine.

Next time it could be you!

Project Kerinci, Sumatra - 2003
In search of the bipedal ape Orang Pendek

How is the Centre for Fortean Zoology funded?

We have no magic sources of income. All our funds come from donations, membership fees, works that we do for TV, radio or magazines, and sales of our publications and merchandise. We are always looking for corporate sponsorship, and other sources of revenue. If you have any ideas for fund-raising please let us know. However, unlike other cryptozoological organisations in the past, we do not live in an intellectual ivory tower. We are not afraid to get our hands dirty, and furthermore we are not one of those organisations where the membership have to raise money so that a privileged few can go on expensive foreign trips. Our research teams both in the UK and abroad, consist of a mixture of experienced and inexperienced personnel. We are truly a community, and work on the premise that the benefits of CFZ membership are open to all.

What do you do with the data you gather from your investigations and expeditions?

Reports of our investigations are published on our website as soon as they are available. Preliminary reports are posted within days of the project finishing.

Each year we publish a 200 page yearbook containing research papers and expedition reports too long to be printed in the journal. We freely circulate our information to anybody who asks for it.

Is the CFZ community purely an electronic one?

No. Each year since 2000 we have held our annual convention - the *Weird Weekend* - in Exeter. It is three days of lectures, workshops, and excursions. But most importantly it is a chance for members of the CFZ to meet each other, and to talk with the members of the permanent directorate in a relaxed and informal setting and preferably with a pint of beer in one hand. Since 2006 - the *Weird Weekend* has been bigger and better and held in the idyllic rural location of Woolsery in North Devon. The 2008 event will be held over the weekend 15-17 August.

Since relocating to North Devon in 2005 we have become ever more closely involved with other community organisations, and we hope that this trend will continue. We also work closely with Police Forces across the UK as consultants for animal mutilation cases, and we intend to forge closer links with the coastguard and other community services. We want to work closely with those who regularly travel into the Bristol Channel, so that if the recent trend of exotic animal visitors to our coastal waters continues, we can be out there as soon as possible.

We are building a Visitor's Centre in rural North Devon. This will not be open to the general public, but will provide a museum, a library and an educational resource for our members (currently over 400) across the globe. We are also planning a youth organisation which will involve children and young people in our activities. We work closely with *Tropiquaria* - a small zoo in north Somerset, and have several exciting conservation projects planned.

Apart from having been the only Fortean Zoological organisation in the world to have consistently published material on all aspects of the subject for over a decade, we have achieved the following concrete results:

- Disproved the myth relating to the headless so-called sea-serpent carcass of Durgan beach in Cornwall 1975
- Disproved the story of the 1988 puma skull of Lustleigh Cleave
- Carried out the only in-depth research ever into the mythos of the Cornish Owlman
- Made the first records of a tropical species of lamprey
- Made the first records of a luminous cave gnat larva in Thailand.
- Discovered a possible new species of British mammal - the beech marten.
- In 1994-6 carried out the first archival fortean zoological survey of Hong Kong.
- In the year 2000, CFZ theories where confirmed when an entirely new species of lizard was found resident in Britain.
- Identified the monster of Martin Mere in Lancashire as a giant wels catfish
- Expanded the known range of Armitage's skink in the Gambia by 80%
- Obtained photographic evidence of the remains of Europe's largest known pike
- Carried out the first ever in-depth study of the *ninki-nanka*
- Carried out the first attempt to breed Puerto Rican cave snails in captivity
- Were the first European explorers to visit the `lost valley` in Sumatra
- Published the first ever evidence for a new tribe of pygmies in Guyana
- Published the first evidence for a new species of caiman in Guyana

EXPEDITIONS & INVESTIGATIONS TO DATE INCLUDE:

- 1998 Puerto Rico, Florida, Mexico *(Chupacabras)*
- 1999 Nevada *(Bigfoot)*
- 2000 Thailand *(Giant snakes called nagas)*
- 2002 Martin Mere *(Giant catfish)*
- 2002 Cleveland *(Wallaby mutilation)*
- 2003 Bolam Lake *(BHM Reports)*
- 2003 Sumatra *(Orang Pendek)*
- 2003 Texas *(Bigfoot; giant snapping turtles)*
- 2004 Sumatra *(Orang Pendek; cigau, a sabre-toothed cat)*
- 2004 Illinois *(Black panthers; cicada swarm)*
- 2004 Texas *(Mystery blue dog)*
- 2004 Puerto Rico *(Chupacabras; carnivorous cave snails)*
- 2005 Belize *(Affiliate expedition for hairy dwarfs)*
- 2005 Mongolia *(Allghoi Khorkhoi aka Mongolian death worm)*
- 2006 Gambia *(Gambo - Gambian sea monster, Ninki Nanka and Armitage's skink)*
- 2006 Llangorse Lake *(Giant pike, giant eels)*
- 2006 Windermere *(Giant eels)*
- 2007 Coniston Water *(Giant eels)*
- 2007 Guyana *(Giant anaconda, didi, water tiger)*

To apply for a <u>FREE</u> information pack about the organisation and details of how to join, plus information on current and future projects, expeditions and events.

Send a stamped and addressed envelope to:

**THE CENTRE FOR FORTEAN ZOOLOGY
MYRTLE COTTAGE, WOOLSERY,
BIDEFORD, NORTH DEVON
EX39 5QR.**

or alternatively visit our website at:
www.cfz.org.uk

Other books available from
CFZ PRESS

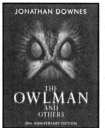

THE OWLMAN AND OTHERS - 30th Anniversary Edition
Jonathan Downes - ISBN 978-1-905723-02-7

£14.99

EASTER 1976 - Two young girls playing in the churchyard of Mawnan Old Church in southern Cornwall were frightened by what they described as a "nasty bird-man". A series of sightings that has continued to the present day. These grotesque and frightening episodes have fascinated researchers for three decades now, and one man has spent years collecting all the available evidence into a book. To mark the 30th anniversary of these sightings, Jonathan Downes has published a special edition of his book.

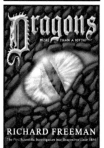

DRAGONS - More than a myth?
Richard Freeman - ISBN 0-9512872-9-X

£14.99

First scientific look at dragons since 1884. It looks at dragon legends worldwide, and examines modern sightings of dragon-like creatures, as well as some of the more esoteric theories surrounding dragonkind.

Dragons are discussed from a folkloric, historical and cryptozoological perspective, and Richard Freeman concludes that: "When your parents told you that dragons don't exist - they lied!"

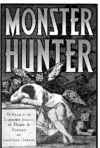

MONSTER HUNTER
Jonathan Downes - ISBN 0-9512872-7-3

£14.99

Jonathan Downes' long-awaited autobiography, *Monster Hunter*...

Written with refreshing candour, it is the extraordinary story of an extraordinary life, in which the author crosses paths with wizards, rock stars, terrorists, and a bewildering array of mythical and not so mythical monsters, and still just about manages to emerge with his sanity intact.......

MONSTER OF THE MERE
Jonathan Downes - ISBN 0-9512872-2-2

£12.50

It all starts on Valentine's Day 2002 when a Lancashire newspaper announces that "Something" has been attacking swans at a nature reserve in Lancashire. Eyewitnesses have reported that a giant unknown creature has been dragging fully grown swans beneath the water at Martin Mere. An intrepid team from the Exeter based Centre for Fortean Zoology, led by the author, make two trips – each of a week – to the lake and its surrounding marshlands. During their investigations they uncover a thrilling and complex web of historical fact and fancy, quasi Fortean occurrences, strange animals and even human sacrifice.

**CFZ PRESS, MYRTLE COTTAGE,
WOOLFARDISWORTHY BIDEFORD,
NORTH DEVON, EX39 5QR
www.cfz.org.uk**

Other books available from
CFZ PRESS

ONLY FOOLS AND GOATSUCKERS
Jonathan Downes - ISBN 0-9512872-3-0

£12.50

In January and February 1998 Jonathan Downes and Graham Inglis of the Centre for Fortean Zoology spent three and a half weeks in Puerto Rico, Mexico and Florida, accompanied by a film crew from UK Channel 4 TV. Their aim was to make a documentary about the terrifying chupacabra - a vampiric creature that exists somewhere in the grey area between folklore and reality. This remarkable book tells the gripping, sometimes scary, and often hilariously funny story of how the boys from the CFZ did their best to subvert the medium of contemporary TV documentary making and actually do their job.

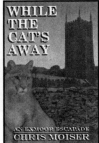

WHILE THE CAT'S AWAY
Chris Moiser - ISBN: 0-9512872-1-4

£7.99

Over the past thirty years or so there have been numerous sightings of large exotic cats, including black leopards, pumas and lynx, in the South West of England. Former Rhodesian soldier Sam McCall moved to North Devon and became a farmer and pub owner when Rhodesia became Zimbabwe in 1980. Over the years despite many of his pub regulars having seen the "Beast of Exmoor" Sam wasn't at all sure that it existed. Then a series of happenings made him change his mind. Chris Moiser—a zoologist—is well known for his research into the mystery cats of the westcountry. This is his first novel.

CFZ EXPEDITION REPORT 2006 - GAMBIA
ISBN 1905723032

£12.50

In July 2006, The J.T.Downes memorial Gambia Expedition - a six-person team - Chris Moiser, Richard Freeman, Chris Clarke, Oll Lewis, Lisa Dowley and Suzi Marsh went to the Gambia, West Africa. They went in search of a dragon-like creature, known to the natives as `Ninki Nanka`, which has terrorized the tiny African state for generations, and has reportedly killed people as recently as the 1990s. They also went to dig up part of a beach where an amateur naturalist claims to have buried the carcass of a mysterious fifteen foot sea monster named 'Gambo', and they sought to find the Armitage's Skink (*Chalcides armitagei*) - a tiny lizard first described in 1922 and only rediscovered in 1989. Here, for the first time, is their story.... With an forward by Dr. Karl Shuker and introduction by Jonathan Downes.

BIG CATS IN BRITAIN YEARBOOK 2006
Edited by Mark Fraser - ISBN 978-1905723-01-7

£10.00

Big cats are said to roam the British Isles and Ireland even now as you are sitting and reading this. People from all walks of life encounter these mysterious felines on a daily basis in every nook and cranny of these two countries. Most are jet-black, some are white, some are brown, in fact big cats of every description and colour are seen by some unsuspecting person while on his or her daily business. 'Big Cats in Britain' are the largest and most active group in the British Isles and Ireland This is their first book. It contains a run-down of every known big cat sighting in the UK during 2005, together with essays by various luminaries of the British big cat research community which place the phenomenon into scientific, cultural, and historical perspective.

CFZ PRESS, MYRTLE COTTAGE,
WOOLSERY, BIDEFORD,
NORTH DEVON, EX39 5QR
w w w . c f z . o r g . u k

Other books available from
CFZ PRESS

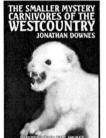

THE SMALLER MYSTERY CARNIVORES OF THE WESTCOUNTRY
Jonathan Downes - ISBN 978-1-905723-05-8

£7.99

Although much has been written in recent years about the mystery big cats which have been reported stalking Westcountry moorlands, little has been written on the subject of the smaller British mystery carnivores. This unique book redresses the balance and examines the current status in the Westcountry of three species thought to be extinct: the Wildcat, the Pine Marten and the Polecat, finding that the truth is far more exciting than the currently held scientific dogma. This book also uncovers evidence suggesting that even more exotic species of small mammal may lurk hitherto unsuspected in the countryside of Devon, Cornwall, Somerset and Dorset.

THE BLACKDOWN MYSTERY
Jonathan Downes - ISBN 978-1-905723-00-3

£7.99

Intrepid members of the CFZ are up to the challenge, and manage to entangle themselves thoroughly in the bizarre trappings of this case. This is the soft underbelly of ufology, rife with unsavoury characters, plenty of drugs and booze." That sums it up quite well, we think. A new edition of the classic 1999 book by legendary fortean author Jonathan Downes. In this remarkable book, Jon weaves a complex tale of conspiracy, anti-conspiracy, quasi-conspiracy and downright lies surrounding an air-crash and alleged UFO incident in Somerset during 1996. However the story is much stranger than that. This excellent and amusing book lifts the lid off much of contemporary forteana and explains far more than it initially promises.

GRANFER'S BIBLE STORIES
John Downes - ISBN 0-9512872-8-1

£7.99

Bible stories in the Devonshire vernacular, each story being told by an old Devon Grandfather - 'Granfer'. These stories are now collected together in a remarkable book presenting selected parts of the Bible as one more-or-less continuous tale in short 'bite sized' stories intended for dipping into or even for bed-time reading. `Granfer` treats the biblical characters as if they were simple country folk living in the next village. Many of the stories are treated with a degree of bucolic humour and kindly irreverence, which not only gives the reader an opportunity to re-evaluate familiar tales in a new light, but do so in both an entertaining and a spiritually uplifting manner.

FRAGRANT HARBOURS DISTANT RIVERS
John Downes - ISBN 0-9512872-5-7

£12.50

Many excellent books have been written about Africa during the second half of the 19th Century, but this one is unique in that it presents the stories of a dozen different people, whose interlinked lives and achievements have as many nuances as any contemporary soap opera. It explains how the events in China and Hong Kong which surrounded the Opium Wars, intimately effected the events in Africa which take up the majority of this book. The author served in the Colonial Service in Nigeria and Hong Kong, during which he found himself following in the footsteps of one of the main characters in this book; Frederick Lugard – the architect of modern Nigeria.

**CFZ PRESS, MYRTLE COTTAGE,
WOOLFARDISWORTHY BIDEFORD,
NORTH DEVON, EX39 5QR
w w w . c f z . o r g . u k**

Other books available from
CFZ PRESS

ANIMALS & MEN - Issues 1 - 5 - In the Beginning
Edited by Jonathan Downes - ISBN 0-9512872-6-5

£12.50

At the beginning of the 21st Century monsters still roam the remote, and sometimes not so remote, corners of our planet. It is our job to search for them. The Centre for Fortean Zoology [CFZ] is the only professional, scientific and full-time organisation in the world dedicated to cryptozoology - the study of unknown animals. Since 1992 the CFZ has carried out an unparalleled programme of research and investigation all over the world. We have carried out expeditions to Sumatra (2003 and 2004), Mongolia (2005), Puerto Rico (1998 and 2004), Mexico (1998), Thailand (2000), Florida (1998), Nevada (1999 and 2003), Texas (2003 and 2004), and Illinois (2004). An introductory essay by Jonathan Downes, notes putting each issue into a historical perspective, and a history of the CFZ.

ANIMALS & MEN - Issues 6 - 10 - The Number of the Beast
Edited by Jonathan Downes - ISBN 978-1-905723-06-5

£12.50

At the beginning of the 21st Century monsters still roam the remote, and sometimes not so remote, corners of our planet. It is our job to search for them. The Centre for Fortean Zoology [CFZ] is the only professional, scientific and full-time organisation in the world dedicated to cryptozoology - the study of unknown animals. Since 1992 the CFZ has carried out an unparalleled programme of research and investigation all over the world. We have carried out expeditions to Sumatra (2003 and 2004), Mongolia (2005), Puerto Rico (1998 and 2004), Mexico (1998), Thailand (2000), Florida (1998), Nevada (1999 and 2003), Texas (2003 and 2004), and Illinois (2004). Preface by Mark North and an introductory essay by Jonathan Downes, notes putting each issue into a historical perspective, and a history of the CFZ.

BIG BIRD! Modern Sightings of Flying Monsters

Ken Gerhard - ISBN 978-1-905723-08-9

£7.99

From all over the dusty U.S./Mexican border come hair-raising stories of modern day encounters with winged monsters of immense size and terrifying appearance. Further field sightings of similar creatures are recorded from all around the globe. What lies behind these weird tales? Ken Gerhard is a native Texan, he lives in the homeland of the monster some call 'Big Bird'. Ken's scholarly work is the first of its kind. On the track of the monster, Ken uncovers cases of animal mutilations, attacks on humans and mounting evidence of a stunning zoological discovery ignored by mainstream science. Keep watching the skies!

STRENGTH THROUGH KOI
They saved Hitler's Koi and other stories

£7.99

Jonathan Downes - ISBN 978-1-905723-04-1

Strength through Koi is a book of short stories - some of them true, some of them less so - by noted cryptozoologist and raconteur Jonathan Downes. The stories are all about koi carp, and their interaction with bigfoot, UFOs, and Nazis. Even the late George Harrison makes an appearance. Very funny in parts, this book is highly recommended for anyone with even a passing interest in aquaculture, but should be taken definitely *cum grano salis*.

**CFZ PRESS, MYRTLE COTTAGE,
WOOLSERY, BIDEFORD,
NORTH DEVON, EX39 5QR**

Other books available from
CFZ PRESS

BIG CATS IN BRITAIN YEARBOOK 2007
Edited by Mark Fraser - ISBN 978-1-905723-09-6

£12.50

People from all walks of life encounter mysterious felids on a daily basis, in every nook and cranny of the UK. Most are jet-black, some are white, some are brown; big cats of every description and colour are seen by some unsuspecting person while on his or her daily business. 'Big Cats in Britain' are the largest and most active research group in the British Isles and Ireland. This book contains a run-down of every known big cat sighting in the UK during 2006, together with essays by various luminaries of the British big cat research community.

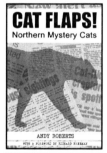

CAT FLAPS! Northern Mystery Cats
Andy Roberts - ISBN 978-1-905723-11-9

£6.99

Of all Britain's mystery beasts, the alien big cats are the most renowned. In recent years the notoriety of these uncatchable, out-of-place predators have eclipsed even the Loch Ness Monster. They slink from the shadows to terrorise a community, and then, as often as not, vanish like ghosts. But now film, photographs, livestock kills, and paw prints show that we can no longer deny the existence of these once-legendary beasts. Here then is a case-study, a true lost classic of Fortean research by one of the country's most respected researchers.

CENTRE FOR FORTEAN ZOOLOGY 2007 YEARBOOK
Edited by Jonathan Downes and Richard Freeman
ISBN 978-1-905723-14-0

£12.50

The Centre For Fortean Zoology Yearbook is a collection of papers and essays too long and detailed for publication in the CFZ Journal *Animals & Men*. With contributions from both well-known researchers, and relative newcomers to the field, the Yearbook provides a forum where new theories can be expounded, and work on little-known cryptids discussed.

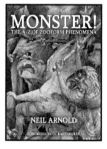

MONSTER! THE A-Z OF ZOOFORM PHENOMENA
Neil Arnold - ISBN 978-1-905723-10-2

£14.99

Zooform Phenomena are the most elusive, and least understood, mystery 'animals'. Indeed, they are not animals at all, and are not even animate in the accepted terms of the word. Author and researcher Neil Arnold is to be commended for a groundbreaking piece of work, and has provided the world's first alphabetical listing of zooforms from around the world.

**CFZ PRESS, MYRTLE COTTAGE,
WOOLFARDISWORTHY BIDEFORD,
NORTH DEVON, EX39 5QR**
www.cfz.org.uk

Other books available from
CFZ PRESS

BIG CATS LOOSE IN BRITAIN
Marcus Matthews - ISBN 978-1-905723-12-6

£14.99

Big Cats: Loose in Britain, looks at the body of anecdotal evidence for such creatures: sightings, livestock kills, paw-prints and photographs, and seeks to determine underlying commonalities and threads of evidence. These two strands are repeatedly woven together into a highly readable, yet scientifically compelling, overview of the big cat phenomenon in Britain.

DARK DORSET
TALES OF MYSTERY, WONDER AND TERROR
Robert. J. Newland and Mark. J. North
ISBN 978-1-905723-15-6

£12.50

This extensively illustrated compendium has over 400 tales and references, making this book by far one of the best in its field. Dark Dorset has been thoroughly researched, and includes many new entries and up to date information never before published. The title of the book speaks for itself, and is indeed not for the faint hearted or those easily shocked.

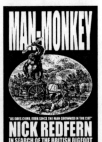

MAN-MONKEY - IN SEARCH OF THE BRITISH BIGFOOT
Nick Redfern - ISBN 978-1-905723-16-4

£9.99

In her 1883 book, *Shropshire Folklore*, Charlotte S. Burne wrote: *'Just before he reached the canal bridge, a strange black creature with great white eyes sprang out of the plantation by the roadside and alighted on his horse's back'*. The creature duly became known as the `Man-Monkey`.

Between 1986 and early 2001, Nick Redfern delved deeply into the mystery of the strange creature of that dark stretch of canal. Now, published for the very first time, are Nick's original interview notes, his files and discoveries; as well as his theories pertaining to what lies at the heart of this diabolical legend.

EXTRAORDINARY ANIMALS REVISITED
Dr Karl Shuker - ISBN 978-1905723171

£14.99

This delightful book is the long-awaited, greatly-expanded new edition of one of Dr Karl Shuker's much-loved early volumes, *Extraordinary Animals Worldwide*. It is a fascinating celebration of what used to be called romantic natural history, examining a dazzling diversity of animal anomalies, creatures of cryptozoology, and all manner of other thought-provoking zoological revelations and continuing controversies down through the ages of wildlife discovery.

**CFZ PRESS, MYRTLE COTTAGE,
WOOLFARDISWORTHY BIDEFORD,
NORTH DEVON, EX39 5QR
www.cfz.org.uk**

Other books available from
CFZ PRESS

DARK DORSET CALENDAR CUSTOMS
Robert J Newland - ISBN 978-1-905723-18-8

£12.50

Much of the intrinsic charm of Dorset folklore is owed to the importance of folk customs. Today only a small amount of these curious and occasionally eccentric customs have survived, while those that still continue have, for many of us, lost their original significance. Why do we eat pancakes on Shrove Tuesday? Why do children dance around the maypole on May Day? Why do we carve pumpkin lanterns at Hallowe'en? All the answers are here! Robert has made an in-depth study of the Dorset country calendar identifying the major feast-days, holidays and celebrations when traditionally such folk customs are practiced.

CENTRE FOR FORTEAN ZOOLOGY 2004 YEARBOOK
Edited by Jonathan Downes and Richard Freeman
ISBN 978-1-905723-14-0

£12.50

The Centre For Fortean Zoology Yearbook is a collection of papers and essays too long and detailed for publication in the CFZ Journal *Animals & Men*. With contributions from both well-known researchers, and relative newcomers to the field, the Yearbook provides a forum where new theories can be expounded, and work on little-known cryptids discussed.

CENTRE FOR FORTEAN ZOOLOGY 2008 YEARBOOK
Edited by Jonathan Downes and Corinna Downes
ISBN 978 -1-905723-19-5

£12.50

The Centre For Fortean Zoology Yearbook is a collection of papers and essays too long and detailed for publication in the CFZ Journal *Animals & Men*. With contributions from both well-known researchers, and relative newcomers to the field, the Yearbook provides a forum where new theories can be expounded, and work on little-known cryptids discussed.

ETHNA'S JOURNAL
Corinna Newton Downes
ISBN 978 -1-905723-21-8

£9.99

Ethna's Journal tells the story of a few months in an alternate Dark Ages, seen through the eyes of Ethna, daughter of Lord Edric. She is an unsophisticated girl from the fortress town of Cragnuth, somewhere in the north of England, who reluctantly gets embroiled in a web of treachery, sorcery and bloody war...

**CFZ PRESS, MYRTLE COTTAGE,
WOOLFARDISWORTHY BIDEFORD,
NORTH DEVON, EX39 5QR
w w w . c f z . o r g . u k**

Other books available from
CFZ PRESS

ANIMALS & MEN - Issues 11 - 15 - The Call of the Wild
Jonathan Downes (Ed) - ISBN 978-1-905723-07-2

£12.50

Since 1994 we have been publishing the world's only dedicated cryptozoology magazine, *Animals & Men*. This volume contains fascimile reprints of issues 11 to 15 and includes articles covering out of place walruses, feathered dinosaurs, possible North American ground sloth survival, the theory of initial bipedalism, mystery whales, mitten crabs in Britain, Barbary lions, out of place animals in Germany, mystery pangolins, the barking beast of Bath, Yorkshire ABCs, Molly the singing oyster, singing mice, the dragons of Yorkshire, singing mice, the bigfoot murders, waspman, British beavers, the migo, Nessie, the weird warbling whatsit of the westcountry, the quagga project and much more...

IN THE WAKE OF BERNARD HEUVELMANS
Michael A Woodley - ISBN 978-1-905723-20-1

£9.99

Everyone is familiar with the nautical maps from the middle ages that were liberally festooned with images of exotic and monstrous animals, but the truth of the matter is that the *idea* of the sea monster is probably as old as humankind itself.

For two hundred years, scientists have been producing speculative classifications of sea serpents, attempting to place them within a zoological framework. This book looks at these successive classification models, and using a new formula produces a sea serpent classification for the 21st Century.

CENTRE FOR FORTEAN ZOOLOGY 1999 YEARBOOK
Edited by Jonathan Downes
ISBN 978 -1-905723-24-9

£12.50

The Centre For Fortean Zoology Yearbook is a collection of papers and essays too long and detailed for publication in the CFZ Journal *Animals & Men*. With contributions from both well-known researchers, and relative newcomers to the field, the Yearbook provides a forum where new theories can be expounded, and work on little-known cryptids discussed.

CENTRE FOR FORTEAN ZOOLOGY 1996 YEARBOOK
Edited by Jonathan Downes
ISBN 978 -1-905723-22-5

£12.50

The Centre For Fortean Zoology Yearbook is a collection of papers and essays too long and detailed for publication in the CFZ Journal *Animals & Men*. With contributions from both well-known researchers, and relative newcomers to the field, the Yearbook provides a forum where new theories can be expounded, and work on little-known cryptids discussed.

**CFZ PRESS, MYRTLE COTTAGE,
WOOLFARDISWORTHY BIDEFORD,
NORTH DEVON, EX39 5QR
www.cfz.org.uk**

Other books available from
CFZ PRESS

BIG CATS IN BRITAIN YEARBOOK 2008
Edited by Mark Fraser - ISBN 978-1-905723-23-2

£12.50

People from all walks of life encounter mysterious felids on a daily basis, in every nook and cranny of the UK. Most are jet-black, some are white, some are brown; big cats of every description and colour are seen by some unsuspecting person while on his or her daily business. 'Big Cats in Britain' are the largest and most active research group in the British Isles and Ireland. This book contains a run-down of every known big cat sighting in the UK during 2007, together with essays by various luminaries of the British big cat research community.

CFZ EXPEDITION REPORT 2007 - GUYANA
ISBN 978-1-905723-25-6

£12.50

Since 1992, the CFZ has carried out an unparalleled programme of research and investigation all over the world. In November 2007, a five-person team - Richard Freeman, Chris Clarke, Paul Rose, Lisa Dowley and Jon Hare went to Guyana, South America. They went in search of giant anacondas, the bigfoot-like didi, and the terrifying water tiger.

Here, for the first time, is their story...With an introduction by Jonathan Downes and forward by Dr. Karl Shuker.

CENTRE FOR FORTEAN ZOOLOGY 2003 YEARBOOK
Edited by Jonathan Downes and Richard Freeman
ISBN 978-1-905723-19-5

£12.50

The Centre For Fortean Zoology Yearbook is a collection of papers and essays too long and detailed for publication in the CFZ Journal *Animals & Men*. With contributions from both well-known researchers, and relative newcomers to the field, the Yearbook provides a forum where new theories can be expounded, and work on little-known cryptids discussed.

CENTRE FOR FORTEAN ZOOLOGY 1997 YEARBOOK
Edited by Jonathan Downes and Graham Inglis
ISBN 978-1-905723-27-0

£12.50

The Centre For Fortean Zoology Yearbook is a collection of papers and essays too long and detailed for publication in the CFZ Journal *Animals & Men*. With contributions from both well-known researchers, and relative newcomers to the field, the Yearbook provides a forum where new theories can be expounded, and work on little-known cryptids discussed.

**CFZ PRESS, MYRTLE COTTAGE,
WOOLFARDISWORTHY BIDEFORD,
NORTH DEVON, EX39 5QR
w w w . c f z . o r g . u k**

Other books available from
CFZ PRESS

CENTRE FOR FORTEAN ZOOLOGY 2000-1 YEARBOOK
Edited by Jonathan Downes and Richard Freeman
ISBN 978-1-905723-19-5

£12.50

The Centre For Fortean Zoology Yearbook is a collection of papers and essays too long and detailed for publication in the CFZ Journal *Animals & Men*. With contributions from both well-known researchers, and relative newcomers to the field, the Yearbook provides a forum where new theories can be expounded, and work on little-known cryptids discussed.

CENTRE FOR FORTEAN ZOOLOGY 2002 YEARBOOK
Edited by Jonathan Downes and Richard Freeman
ISBN 978-1-905723-30-0

£12.50

The Centre For Fortean Zoology Yearbook is a collection of papers and essays too long and detailed for publication in the CFZ Journal *Animals & Men*. With contributions from both well-known researchers, and relative newcomers to the field, the Yearbook provides a forum where new theories can be expounded, and work on little-known cryptids discussed.

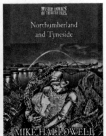

THE MYSTERY ANIMALS OF THE BRITISH ISLES: Northumberland and Tyneside
By Mike Hallowell ISBN 978-1-905723-29-4

£14.99

This is the first volume in a major new series from CFZ Press, which attempts nothing less than chronicling all the mystery animals, zooform phenomena, aberrations, and animal folklore of the United Kingdom and the Republic of Ireland. In this volume Mike introduces us to sea serpents, dragons, giant lobsters, ghost birds, the ghost of *Wandering Willie*, and even a vampire rabbit. A fantrastic book and a great introduction to the series.

CENTRE FOR FORTEAN ZOOLOGY 1997 YEARBOOK
Edited by Jonathan Downes and Richard Freeman
ISBN 978-1-905723-31-7

The Centre For Fortean Zoology Yearbook is a collection of papers and essays too long and detailed for publication in the CFZ Journal *Animals & Men*. With contributions from both well-known researchers, and relative newcomers to the field, the Yearbook provides a forum where new theories can be expounded, and work on little-known cryptids discussed.

**CFZ PRESS, MYRTLE COTTAGE,
WOOLFARDISWORTHY BIDEFORD,
NORTH DEVON, EX39 5QR
www.cfz.org.uk**

Printed in the United Kingdom by
Lightning Source UK Ltd., Milton Keynes
137325UK00001B/92/P